PRAISE FOR DEBORAH BRECHER'S METHOD:

"Highly recommended for the person who wants to learn about computers but doesn't know where to start. A practical course taught in layman's language."

—Barbara Le Compte,
Staff Assistant,
Systems Control Technology

"Since taking the class, I've started to use a computer at work. You took the mystery out of it."

—Annette Campbell, RN,
Pacific Medical Center

"I attribute my employment as a technical writer to the literacy course. It was that class that gave me critical knowledge about computers."

—Sheri Hartman, Writer,
ASK Computer Systems

"All the mystery has been taken away and for once I can actually say I feel a part of the computer world. It was so simple, with the right technique of instruction."

—Paula G. Evans,
Executive Secretary,
Apple Computer

DEBORAH BRECHER is co-creator of the Women's Computer Literacy Project in San Francisco. This book is the product of her experience with the Project as well as her constant communication with other women's professional groups.

THE
WOMEN'S
COMPUTER LITERACY
HANDBOOK

Deborah L. Brecher

A PLUME BOOK

NEW AMERICAN LIBRARY

A DIVISION OF PENGUIN BOOKS USA INC., NEW YORK

PUBLISHED IN CANADA BY
PENGUIN BOOKS CANADA LIMITED, MARKHAM, ONTARIO

NAL BOOKS ARE AVAILABLE AT QUANTITY DISCOUNTS WHEN USED TO
PROMOTE PRODUCTS OR SERVICES. FOR INFORMATION PLEASE WRITE
TO PREMIUM MARKETING DIVISION, NEW AMERICAN LIBRARY,
1633 BROADWAY, NEW YORK, NEW YORK 10019.

PLUME TRADEMARK REG. U.S. PAT. OFF. AND FOREIGN COUNTRIES
REG. TRADEMARK—MARCA REGISTRADA
HECHO EN HARRISONBURG, VA., U.S.A.

SIGNET, SIGNET CLASSIC, MENTOR, ONYX, PLUME, MERIDIAN and NAL
BOOKS are published *in the United States* by New American Library,
a division of Penguin Books USA Inc., 1633 Broadway, New York, New York
10019, and *in Canada* by Penguin Books Canada Limited, 2801 John Street,
Markham, Ontario L3R 1B4

Library of Congress Cataloging in Publication Data

Brecher, Deborah L.
    The women's computer literacy handbook.

    Includes index.
    1. Computer literacy.   I. Title.
QA76.9.C64B74   1985   001.64   85-7087
ISBN 0-452-25565-1

First Printing, August, 1985

    5 6 7 8 9 10 11

PRINTED IN THE UNITED STATES OF AMERICA

To Ada Lovelace,

the world's first programmer. The story of Ada Lovelace is one more example of how women's achievements get "lost." Ada Augusta (1815–1852) was Lord Byron's daughter and later became Countess Lovelace. After meeting Charles Babbage, inventor of the first computer, she became interested in the practical aspects of getting the machine to actually solve problems. As the first programmer, she realized that the binary system should be used in computer mathematics instead of the decimal system used by Babbage. Although Charles Babbage's accomplishments are recognized and discussed in most introductory computer science textbooks, Ada Lovelace's pioneering work is rarely if ever mentioned.

Acknowledgments

This handbook owes its existence to many, many people. First and foremost is Jill Lippitt, my partner, who read and reread the manuscript, editing each version to keep me from slipping into the jargon I have tried so hard to avoid. Billie Miracle, my illustrator, was a delight to work with. She effortlessly transformed my mental visualizations into concrete images. And, of course, this book would never have come into being without the creation of our school, the Women's Computer Literacy Project, and the wonderful organizing talents of Marcia Freedmann, one of the Project's co-founders.

Although the need for the curriculuum was obvious, neither the school nor the book might exist if not for the encouragement and support we received from the American Association of University Women (AAUW) and its national technology chairwoman, Corky Bush. The AAUW was the first women's organization to understand the tremendous impact that computer technology would have on women's lives. It adopted the subject as a study topic while other progressive groups were still hoping that computers would go away.

Una Glass and the staff at her store, Matrix Computers, were always available with help when I needed it the most. Meeting and working with a woman technician (repairperson), Lucinda Dekker, also provided the help I needed to get over my reluctance to actually touch the innards of the computer.

Heidi Key volunteered her time most generously. Her work in tailoring software for our classroom computers was invaluable. Most importantly for my own work, she served as a model for me, always keeping her "cool" when the computers reminded us that they were machines and we were all too human.

Frieda Feen assumed most of the burdens of my other project, the National Women's Mailing List. Her work gave me the free time I needed to devote to this book.

Once computer phobia leaves it is often replaced by the feelings of transference usually reserved for therapists. Needless to say, I have fallen in love with my KayPro computer and the WordStar programs that I used to write and index the book. My thanks to both KayPro and MicroPro for bringing these products into being.

And finally, special thanks to CompuPro Computers for providing us with state-of-the-art computers for our classroom. CompuPro recognized the need for special classes for women at a time when the more famous computer manufacturers were giving us the cold shoulder.

Contents

THE
WOMEN'S
COMPUTER LITERACY
HANDBOOK

Introduction

In the Beginning ...

As this book goes to press, I am celebrating my twentieth anniversary of working with computers. I've done every job in the computer industry—from working as a computer operator to being in charge of 1,000 data processing employees in a government agency, to being president of my own computer consulting firm. I started in the olden days of computing, using equipment that cost over $1 million and took up huge roomfuls of space. Today that same computer power is available in a box that fits on your desk, taking up about the same space as an electric typewriter, and costing under $2,000.

I've seen incredible changes in the computer industry, but one thing has never changed—the frustration and anxiety that surround learning about using these machines. For 20 years the manufacturers have been promising that their machines will act friendlier and be easier to learn to use, but the totally friendly computer still has not been invented.

Most of the suffering that goes along with learning about computers is not really necessary. In large part the problem lies with the ways computing is taught and with the instruction manuals, rather than with computers themselves. My personal experience has been that computers are easy to work with (and fun) once you break through the mystique that surrounds them. This is not to say that I didn't shed my share of tears of frustration while learning—I most certainly did. But when comprehension finally dawned, I discovered that the concepts I was learning were easy—and they were

1

similar to what I already understood. The problem was that no one was teaching computers in terms of things I understood.

As my career progressed, I realized that the mystique that surrounds the computer industry was actually quite useful to the cognoscenti—those in the know. Computer professionals had a vested interest in keeping computers hard to understand and work with. It's called job security, and it not only resulted in job security but high pay as well.

Naturally, I enjoyed the benefits of being a computer professional myself. Not only did my mother think I was brilliant, but my employers didn't complain about the work schedule I set for myself, or what I wore, as long as I kept their computer systems working. But at a certain point those rewards began to wear thin. I decided to do something more socially meaningful with my skills. I founded the National Women's Mailing List as a computer-based communications and networking resource. This project uses a personal computer and a packaged computer program that was available off the shelf. It allows over 60,000 women to receive mail about any number of women's issues that they have checked off on a registration form. My partner, who had no previous computer experience, was quickly able to operate the system, using only the literacy skills that are covered in this handbook. She was the first of my many successful students.

In the summer of 1982, I went on a cross-country tour with a computer packed carefully in the trunk of my car. I met with hundreds of representatives from women's organizations to show them firsthand what computers could do to help them in their work. In each of the 13 cities that I visited, women had the same reaction. "Great! But how do I begin if I don't know anything about computers?"

To answer that question, I investigated the classes and books that were available and realized that there really wasn't any good beginning place to learn about computers in general. Particularly, there was nothing available that made it easy for women to gain these skills without being patronized, put down, or paralyzed by unnecessary fears. After being repeatedly requested to teach women how to get started on computers, I developed an in-depth, straightforward, 2-day curriculum and opened a computer school for women in San Francisco. This book is based on the curriculum of that school—The Women's Computer Literacy Project.

The Literacy Project has been overwhelmingly successful, and we now teach classes regularly in New York, San Francisco, and Anchorage, as well as special courses throughout the country. In all these locations, the reactions from our students have been remarkably consistent. It seems that no matter how little or how much they already know about computers, they report that the class is exactly what they need!

Who Should Read This Book?

Women Who Know Nothing about Computers (but know it's time to learn)

What's so remarkable about my students' reactions is that in any one class there is an incredibly diverse group of women with varying computer experience, including some who know absolutely nothing about computers. For them my curriculum is perfect because I never use any "computerese" without first explaining the term. Amazingly, this is not true of the majority of books and classes.

One student was an elementary school teacher. She knew that many of her third graders were using computers at home and she wanted to catch up to them. In particular, she wanted to begin thinking about how she could integrate computers into her classroom when they arrived, which she knew was only a matter of time. So, this teacher took a beginning class at the local community college—a class that was advertised as especially for public school teachers. What happened? She had a disastrous experience and dropped out because the teacher used so much computer jaragon that she couldn't figure out what he was talking about. Since a good number of the students already knew the jargon (or pretended they did), she felt she was holding up the class if she interrupted to get the terms explained. Even then, the teacher had a hard time putting computerese into words she could understand and usually explained these terms by using even more technical words, which, of course, had not been defined either. In my class she caught on quickly, had fun, and learned what she needed to know.

In both the Literacy Project's classroom and in this *Computer Literacy Handbook*, you'll find one rule followed throughout—no technical words are used without first being defined. A second rule I follow is that all definitions are based on things you already know.

These rules are just what any good teacher knows—real education occurs best when learning is based on the student's own experiences. Instead of expecting people to memorize confusing definitions blindly, it is more effective to present new concepts by referring to things with which the student is already familiar. Therefore, all the computer concepts in this book are introduced by using analogies to things you already know. This approach results in the new material being easy. That's one of the big surprises for my students—computers are easy! That really shouldn't be a great shock to you. Remember, 12-year-olds do quite well with computers, and anything an average 12-year-old can do can't be that difficult.

Women with Some Familiarity with the Vocabulary (but who haven't been able to put it all together yet)

Another typical student in our classes is the woman who has been talking about computers for quite some time. She may be an account executive or copywriter at an advertising agency working on a computer account, or she may be a middle-management executive in a corporation that is in the midst of computerizing. Perhaps she is the wife of a programmer and has to endure endless discussions of software. Or she may be the mother of a 12-year-old who has become obsessed with her Atari. She may even be the supervisor of workers who use computers, although she doesn't use them herself. In any case, she has finally gotten tired of having to bluff her way through, without really understanding how all the separate pieces of knowledge that she's acquired fit together.

The extent of this problem is typified by one of my students who is the vice-president of a midsize computer company. She took the class rather anonymously, but she confided to me afterward that it would be a relief not to have to continue bluffing about so many vocabulary words and concepts anymore. Although she had advanced to a top position in the industry, she had been unable to really put it all together on her own.

And that's what this book does—it puts it all together by integrating the terms and concepts of computers into one manageable handbook. The benefits from understanding computers are immense. You will have the confidence to purchase the computer that is right for you. If you already have a computer at work or at home, you'll be able to extend what you already know to get the most out of it. All of a sudden you will be able to read the computer ads and understand what's being sold. And when there are reports of new breakthroughs in computer technology or when new products such as the AT&T personal computer are introduced, you'll be able to appreciate their significance. In short, you'll be a full-fledged member of the twentieth century, able to understand the world around you.

Understanding the world of computers is becoming mandatory. One of my students reported excitedly that all of a sudden cocktail parties were more enjoyable because she understood all those "bits" and "bytes" everyone was talking about. That may not be of such crucial importance to you, but what about the feeling you have when you try to take a telephone message for your co-worker and feel like the person on the other end is speaking Greek? I once got a phone message from my secretary about someone wanting to borrow my "300 Bog Odem." It wasn't until I returned the call that I learned that the request was really for my 300 baud modem. (After reading this book, you'll understand what this is, as does my secretary, who has now taken my class.)

Women Who Already Use Computers (but want to do more)

In our classes, we always have some students who already use computers. For these students, the computer isn't intimidating or hateful. These women have already made the discovery that the computer is helpful and even fun. But they want to learn and do more. Most of these students have already realized that the instruction manuals that came with their computers are no help but instead are obstacles to overcome.

After they receive training in computer literacy, students are able to read those instruction manuals that once seemed so mystifying. And no one has to teach these women how to use the programs that came with their computer. They can teach themselves because they are now computer literate.

It's exciting to watch secretaries realize how much they already know about computers. One of the sinister aspects of office technology is that the technical terms that secretaries learn to do word processing are different from the vocabulary used for the same concepts and procedures in the rest of the computer industry. Because a word processor is a computer, the secretary really is an experienced computer user, although she often doesn't know it. She doesn't realize that all the seemingly incomprehensible computer terms that she hears are really the same as the word processing jargon she already knows well. Thus, most secretaries have less job mobility than other computer professionals, until they realize just how valuable their skills actually are. Those who do understand the connections between word processors and computers have been able to transfer those skills successfully to other segments of the industry.

My favorite phone calls come from former students who tell me all about what they've gone on to teach themselves on their jobs. They've often mastered systems that I've never used myself. And that's the point of literacy after all. Once you are able to read and understand computer books, you can teach yourself things even your instructor doesn't know.

Commonly Asked Questions

As the Literacy Project's director of education, I have been interviewed by reporters for television, radio, and the print media. The questions that they ask are remarkably consistent. "What is computer literacy?" "Why have a special *women's* class or a book just for women?" "Don't computers act the same whether a man or woman is using them?" "Why isn't there a *man's* book of computer literacy?" If you're wondering if this is the right book for you, I imagine you're asking yourself those questions, too. Therefore, answering them is a good place to begin.

What Is Computer Literacy?

First, let's look at literacy in general. Literacy simply means you can read (and understand what you read). Computer literacy is similar—it means you can read (and understand) books about computers. One of the objectives of this book is to give you the ability to read other computer books and manuals. The whole notion of needing a book to teach you how to read other books may seem odd. But have you looked at a computer manual lately? I've had 20 years of experience with computers and most of the so-called introductory books either give me a headache or make me want to scream.

What takes getting used to is that there is a computer term for everything. Computer vocabulary creates a barrier to understanding. Until you understand that vocabulary, you simply can't read the material about computers. The existence of a confusing vocabulary is no accident. The phenomenon of everyday people being unable to read instruction manuals creates the need for an elite class of workers to minister to the needs of those unfortunates who are unable to manage on their own. So a profession is created to fill this need—highly paid computer professionals.

Every profession has its own vocabulary. But the computer profession has outdone all others. For example, computer instruction manuals aren't really called instruction manuals. They're called **documentation**. Why have a special word for instruction manuals? Does it add any preciseness of understanding? What it does do is ensure that if you are a technical writer writing documentation, you earn more than a technical writer writing instruction manuals about noncomputer subjects. That's one of the main objectives in creating an elite profession—earning more money.

To be fair, in many cases a specialized vocabulary has arisen to save time and space. That is, a single word was created to convey a special process or as shorthand for a common concept. When you read a cookbook, for example, you encounter cooking vocabulary. Words such as "sauté," "baste," and "parboil" save time and space. However, in computing, there are existing words that could easily have been used. For example, you have probably heard of the word **byte**. It simply means a character, that is, a letter, number or special character. (The letters "a" and "b"; the numbers "1" and "2"; and the special characters "$" and "%" are examples of characters.) The word "byte" does not really convey any more meaning than the perfectly good word "character," but it does sound mysterious.[1]

Each brand of computer operates differently. No one, not even a Ph.D. in computer science, can operate a computer system she hasn't used before without first reading the instruction manual. Unfortunately, these manuals are written in computerese, and they assume the reader is already familiar with the computer words and concepts they use. But unlike other technical manuals—for example, automobile owners' handbooks, which many owners never read—everyone who uses a computer has to use their manuals on a regular basis. When things go wrong, the salesperson at the computer store won't be the one to help you—you'll be told to read the documentation that came with your system. Some computer systems give you a help service—an 800 telephone number that you call when you have questions. But if you can't communicate in the language of the technician at the other end, don't expect to get much help.

Once you can read computer manuals with understanding, the world of information processing is at your command. You can teach yourself how to use your husband's or child's home computer or the computer at work. You can shop for a computer that meets your particular needs and be able to tell the differences among the various models. You can select the specific kinds of computer programs you will need and know how to make sure they will do what you want. You will be able to rely on your own judgment rather than on the (mis)information of computer salespeople. Then, once you have bought a system, you will be able to sit down and follow the operating instructions to make it work.

Why Should Women Learn about New Technology?

The most obvious reason for learning about technology is to simply have the option of working. It will only be a short time before *all* jobs are computer dependent. Employment specialists refer to this as the coming employment dislocation, when workers find that their existing skills are no longer of any value.

That possibility may seem very remote to you if you are a white-collar worker. However, consider the banking industry. With automated tellers increasing in popularity, the need for human tellers has decreased. It is only a matter of time before human tellers will simply cease to exist. Like blacksmiths and buggy whip manufacturers, they will become superfluous. Have you called the information operator for a telephone number lately? In many regions of the country a computer-generated voice tells you the number. Before long, voice-recognition equipment will completely replace the job of information operators. Experts are now projecting that within the next 5 years every white-collar worker will be using a computer. And whether you realize it or not, your job will be affected too.

It isn't hard to imagine a future in which only people with computer skills will be employable, except in the very lowest-paid sectors. Imagine a group of people that has some discomfort with machines and technology. This group will be the first to feel the effects of new technology—unemployment. Women, in general, tend to back away from technology and electronic machines and, as a result, will suffer the direst economic consequences unless they overcome their "technophobia."

Just as compelling as the need for marketable skills is the importance of having our new electronic age reflect the sensibilities of women. To do so, large numbers of women must enter and advance within the technology fields. In fact, the percentage of female engineers has gone down! The result of this abdication by women is that our increasingly technological world is being shaped and directed almost entirely by men. Most women understand this problem on a somewhat subconscious level. We know that the world in which we live does not generally support or reflect women's priorities. Any working mother who has to struggle with the lack of child care knows firsthand that society does not really support her need to both work and raise a family. I can't help but believe that if there were equal numbers of women and men in the highest levels of government, business, and labor unions, child care centers would be supplied as a matter of course, just as social security is deducted from your paycheck.

Would women as policymakers be any different? I like to think so. I think that women have certain valuable priorities and concerns that grow out of the female experience. Inevitably, if there were equal numbers of men and women policymakers, this unique perspective would be allowed expression and the decisions that were made would include it. These beliefs are reinforced when I think of the important contributions that women have made toward changing the perception of certain social issues. For example, there is Rachel Carson, whose concern for the ecology of the planet virtually started the environmental movement with her book *Silent Spring*. Or Helen Caldicott, one of the first physicians to speak out against the charade of a survivable nuclear war. Or think of premenstrual syndrome (PMS). Thousands of ailing women were ignored by their male doctors until a female medical researcher found the physiological cause of PMS.

Are women computer professionals different? Can they have a salutary effect on the developing technological age? We may never know. Unless women enter the technology fields now, while the new age is still young and the doors are still open, the electronic age will fail to reflect women's sensibilities, just as the industrial age did before it. If women don't catch up soon on basic technical skills, we may lose all the advances within society that we've worked so hard for in the past.

When I bring up this subject—the gender gap around technological subjects—many people assume that things have changed. I wish I could be that optimistic. Teachers tell me that they're seeing the same old patterns. When computers are in the elementary school, they are dominated by boys. Some elementary schools have felt that to get girls to use computers at all, they had to introduce gender-separated learning—boys on one day and girls on another. That way the girls didn't have to compete with boys, who are generally more aggressive around the equipment.

High school teachers report that being good at computers is not a popular thing for girls. What interest the girls have in computer technology drops off markedly at puberty when they become romantically interested in boys. It should be no surprise that being popular and having dates is more important than knowing about computers for high school girls. They are reluctant to hurt their image by pursuing technological interests. Although it may not be seen as a macho thing to be good at computers, it still isn't considered feminine.

What Is the Woman's Approach to Understanding Computers?

Learning about computers was a horribly frustrating experience for me. My education was less than a success when my male co-workers showed me what to do and expected me to memorize their rote instructions. I found that in order to understand computer concepts I first needed to have a holistic sense of how the machine worked and what was going on inside of it. Only after I figured out the reason behind the rules did I understand, and truly master, the computer.

In my work as director of education for the Women's Computer Literacy Project, I have taught hundreds of women who, like myself, were unable to feel comfortable operating a computer unless they knew what they were doing and, most importantly, *why*, rather than blindly following instructions. Without this kind of understanding, most women do not have the courage to touch a computer, even if they have been shown how, for fear of breaking it. Men, however, seem braver about playing around with the machine and learning by trial and error.

A common myth is that you need to be a mathematician or a scientist to understand how computers work. This myth is as dangerous as it is wrong, because it serves to trigger women's traditional math/science phobias, with the result, of course, of keeping women out of the field. In fact, the basic information that you need to know about how a computer works is relatively easy to conceptualize. I will present this information through a series of analogies, drawn from women's everyday experience, that I have successfully developed first for myself and then later for my women students. They are guaranteed to work.

A by-product of this approach—understanding from the inside out—is that you will know as much, if not more, than the salesperson from whom you will be buying your equipment. This is an excellent position to be in, since buying a computer is similar to buying a used car—the more you know, the better your chances of getting a good deal. And like buying a used car, you're taking your chances if you make a decision relying on what the salesperson tells you. Unfortunately, all too often computer salespeople know more about selling than they do about computers.

Why a Woman's Book?

I have called this *The Women's Computer Literacy Handbook* in order to emphasize that the explanations of computer technology reflect my own, very personal, frame of reference. Since I am a woman, the analogies and examples are drawn from the female experience. I have chosen to emphasize this fact in the title because almost all books on technological subjects do the reverse—they base themselves on a male environment. Male-centered books on technology are taken to be the norm, using masculine pronouns and examples drawn primarily from men's experience. Authors never point out that a book is the *Man's Handbook of Mechanics*, for example. They don't have to—male experience is assumed.

I believe that the use of male-centered examples makes it that much harder for women to learn subjects that are themselves thought to be the domain of men. When women do persevere and learn these subjects, they are said to have "nontraditional" skills. What does this say about our society? Again, it emphasizes that our technological world is not being created by a partnership of men and women. Women make up only 2 percent of engineering graduates. Why does engineering (also math, physics, chemistry, automobile mechanics, and plumbing) "turn off" women? I suggest that the problem lies with the methods of teaching. Why does any subject seem hard? When any subject is truly understood, it is easy. But in order to understand new material, it is essential that the subject matter be presented within a framework, or context, that is already familiar. This book recognizes that most computer concepts considered difficult or complex are actually similar to things that women are already familiar with as part of their everyday lives. It is actually simple to grasp technological subjects through an analogy to something you already know. In this book all the examples refer to experiences and objects that most women (and men) are used to; thus, learning about computers becomes easy.

As you will see, this book generally uses female pronouns when referring to people working in technological jobs. This is not done to "turn-off" men or because of female chauvinism. To the contrary, men should read this book, too, if they want computer technology made easy and accessible. I've chosen to use female pronouns to make the point that women can and do work within technology and that there are female role models to inspire us. This may seem like an unimportant detail, but after many years of answering the Male Help Wanted section of the want ads, I was relieved when those classifications were prohibited. Small details do affect us, albeit subconsciously.

Although this is a book written to encourage women, there is a fine line between choosing familiar analogies and condescending to the woman reader. Many of my examples refer to cooking and baking. Those examples were selected because they bear such an amazing resemblance to computer functions. That is quite different from suggesting that you use a computer to store your recipes. It is the latter proposal that smacks of condescension.

When people ask me, "Should I read a woman's book?" what they often mean is that they're afraid that the material will be too simplistic or watered down. On the contrary, this book covers many computer concepts that are rarely discussed in beginners' books because they are believed to be too difficult. The fact is, as I've said before, with the right analogy nothing is too hard.

There is another reason for calling this *The Women's Computer Literacy Handbook*. That is to remind you that the material in this book includes a feminist approach to technology. The key element of this approach is a holistic awareness of the entire system, which includes both the human operator as well as the machine. This approach is particularly suited to any discussion of computers, because computers are changing the way we work and live. Since computers are specialized tools for managing information, we can expect them to affect traditional "women's work" the most. It is the secretaries and file clerks who will feel the greatest effect from the new automated office. It is women who sit at computer screens all day, whether they are working as telephone operators or order-entry clerks at the chain store. These women will feel any health effect of low-level radiation first—not the middle-management executives who made the decision to automate the office. Although the health risks from radiation have not been determined yet, there have been some alarming reports of birth defects related to pregnant women working at computer terminals. Much research still needs to be done in this area. As informed and educated "computer literates," we will be best able to push for such needed research.

Every day we read in the newspapers of the wonderful benefits of our new computer age. But little or no coverage is given to incidents such as the Blue Shield workers' strike. How many have heard of the strike by the claims processors in San Francisco? These women (and they were all women) went on strike for the right to stand up and stretch to rest their eyes from the strain of working at a computer screen all day. They typed at computer terminals and wanted to be able to relax their eyes for 10 minutes every hour. They lost their strike because Blue Cross moved their claims processing to a nonunionized region. The reason that the women had to ask formally for the right to take a break was because the computer was doing **keystroke monitoring**. This means that the computer counted every key that the worker struck. Quotas were set as to the number of hourly keystrokes required. The clerk had to get permission for a break or she would fall below her quota.

This image of the computer monitoring our work has all the signs of a world like Orwell's *1984*. Whether or not that vision of an authoritarian world becomes a reality depends in part upon how computers are used and who is in a position to set technological work policies. Again, it is up to each of us to become informed and knowledgeable about these powerful machines. It is only through such knowledge that we will be able to become informed consumers and workers who can have some effect upon the policy decisions that determine how these incredibly powerful tools are used. Whether we choose to understand these tools or not, they are here to stay, and none of our lives will be unchanged.

Chapter 1

Computers from the Inside Out

In this chapter you will learn what many of those computer buzzwords you've been hearing actually mean. You will find out what the parts of a computer system are and how they all fit together. As the mysteries of this technical jargon are unveiled, you will be surprised to see how easy it really is.

Basic Vocabulary

Let's begin with **hardware.** This word just means the things you can touch. If you can bang on it, it's hardware. Hardware is simply a fancy term for equipment.

The word hardware was invented to differentiate it from **software.** Software is another name for computer programs. We'll talk more about software later. Now, you're probably asking, "But what is a **program?**" I had a hard time understanding what a computer program is until I realized that it's really nothing more than a recipe that the computer can follow.

When you cook, *you* are the computer and the recipe is the program. You can read an instruction and perform it. Then you read the next instruction and perform it. After you've completed the whole set of instructions, you end up with the desired result, a chocolate cake, perhaps. In a similar way the computer follows instructions—one at a time, just like you.

17

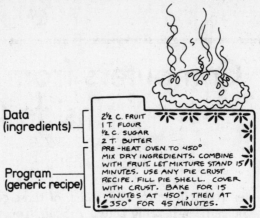

Data
(ingredients)

Program
(generic recipe)

2½ C. FRUIT
1 T. FLOUR
½ C. SUGAR
2 T. BUTTER
PRE-HEAT OVEN TO 450°
MIX DRY INGREDIENTS. COMBINE
WITH FRUIT. LET MIXTURE STAND 15
MINUTES. USE ANY PIE CRUST
RECIPE. FILL PIE SHELL. COVER
WITH CRUST. BAKE FOR 15
MINUTES AT 450°, THEN AT
350° FOR 45 MINUTES.

FIGURE 1-1. A program is a like generic recipe. The procedure section is the program, and the ingredients are the data.

Think about recipes for a moment. They all have a standard format. There are two parts: First, there is a list of ingredients, such as eggs, milk, butter, sugar, flour, and fruit. Then, there are the instructions that actually tell you how to make the dish. These instructions might tell you to blend the milk and flour together or to beat the egg whites.

As Figure 1-1 shows, a computer program is the part of the recipe that sets out the procedures the cook performs. The ingredients are the **data.** That's what's so special about computer programs. They are generic recipes, that is, procedures without the ingredients (data). For example, if you had a generic recipe for a fruit pie, the pie you baked would depend on what kind of fruit you added. Cherries would make a cherry pie, apples would make an apple pie, and so on.

When you buy a program (recipe), you get the set of instructions that the computer will follow. For example, you can buy a program that has instructions for the computer to do payroll accounting. You must then give the program the data (ingredients) that it will use. Data can be the names of your employees, their wage rates, hours they work, and so on. The data will allow the program to deal with your particular payroll. The result of the program depends upon what data you, the **user**, provide.

Here's another example: if you use a mailing list program (a generic recipe), you could type in data that are California names and addresses (ingredients), in which case you'd get a California mailing list. Or you could type in data that are New York names and addresses, in which case you'd get a New York mailing list. Like a cook using a generic recipe, the computer follows a set of instructions. In both cases the finished product depends on the type of ingredients or data, the cook, or computer user, puts into it. This concept of generic recipes may seem unfamiliar because we're used to recipes for a specific dish. Just keep in mind that generic recipes are recipes with instructions but not ingredients.

When a **programmer** writes a program for a computer, it's similar to a baker who invents a new recipe (maybe for a "Pillsbury bake-off"). They must both think about all the necessary steps to accomplish the desired result and must set them out in orderly and clear instructions. If, for example, a programmer was designing a mailing list program (recipe), she would have to figure out exactly how to tell the computer to take the names and addresses (the data), sort them into zip code order, and then print them in the proper way on mailing labels.

Just as most cooks do not create original recipes but use recipes in cookbooks, most computer users do not write programs. Instead, they buy programs that have been written and tested by specialists (programmers).

Now lets get back to software. Exactly why are programs called software? Think about your cassette tape recorder and the prerecorded tapes you buy. You can put the recorder on play and hear the music, but there is no way to actually touch the music. You can touch the tape and you can touch the tape cassette recorder, but the music itself is intangible. You bought it, you can experience it, but there is no way to touch it. Thus, you might say the music is "soft."

Similarly, you can buy a prerecorded computer program

on a cassette tape. The prerecorded tape might contain the program (instructions) to make the computer play a game such as Pac-Man. That is, when the cassette player is plugged into the computer, you can put the Pac-Man tape on the cassette player and press the play button. The computer will receive the instructions and follow them, enabling you to play the game. Just as in our music example, you can physically touch the cassette player, but you can't touch the instructions themselves. The cassette player and the computer are hard or touchable—they're hardware. The program is real but intangible. Like music, it's "soft." That's why computer programs have come to be called software.

What Is a Computer?

"Big deal," you may be saying. "Hardware is equipment, a program is a recipe, and some people call those recipes software. But what is a **computer?** And why are so many people buying them?"

In the old days before computers, most machines could only accomplish one thing. I remember visiting my mother at the insurance company where she worked. In the office there was a billing machine. It was created to do one thing —billing. Special ledger cards were inserted, and payments were mechanically posted to each client's account. This machine saved the bookkeeper a lot of time, but it had a serious drawback—it could not do anything other than billing. You couldn't use it to analyze cash flow projections or to do next year's budgeting.

But computers are very different—they are multipurpose machines. Think about our cooking analogy again. A cook can follow many different recipes, from soups to meats to desserts. In the same way, a computer can run many different types of programs, from word processing, to graphics, to spreadsheets, to accounting (to name a few). When a computer uses a particular program, it "becomes" a specific kind of machine. A single computer can become an adding machine, a billing machine, a budgeting machine, a payroll machine, a word processor (more about that later), a graphics machine, an inventory controller, as well as many other kinds of machines. It all depends on what kind of software it is using.

So what *is* a computer? Science fiction stories make it seem like an incredibly powerful machine. I find it reassuring to know that it can really only do two things: It can add, and it can compare.[2]

Comparing means it can take two things, see if they are greater than, less than, or equal to each other and, based upon the result of the comparison, take appropriate action. That sounds complicated, but it's what your thermostat does.

If you set the thermostat needle to 70°, the thermostat compares the room temperature to the needle setting. When the room temperature is 68°, the result of the comparison is that the room temperature is less than the the needle setting, and the appropriate action is to turn on the heater. When the room temperature reaches 71°, the comparison shows that the room temperature is greater than the setting of the thermostat, and the appropriate action is to turn off the heater.

We are all familiar with the computer's other ability, adding. It's what your pocket calculator does. But, it is the ability to combine these elementary capabilities that gives the computer the appearance of having other, more sophisticated abilities. This is what often passes as computer "intelligence."

Let's look at your pocket calculator again, which, after all, is a little computer. Suppose you want to multiply two numbers, for example, 23 by 4. What really happens is that the calculator has to add 23 four times. After all, it doesn't know how to multiply, it only appears to have that ability because it follows the instructions to add so fast. When you press the multiplication key, the calculator begins to follow a set of programmed instructions. To understand this, look at Figure 1-2 as you read on.

The calculator starts out with 0 and adds the first number you gave it—23. Then it adds 1 to a **counter.** (Think of a counter as the clicker that ticket takers use to count how many people are going in to see a show. Every time the

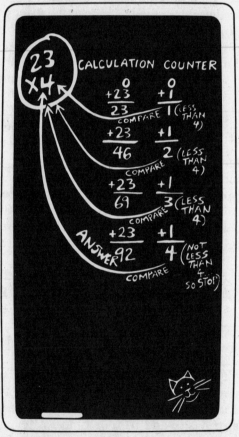

FIGURE 1-2. A computer can only add or compare, but it appears to have other capabilities such as multiplication. However, these abilities are just the result of repeatedly adding and comparing.

clicker is pressed, 1 is added to the total.) Then the calculator compares the value in the counter to the second number
you gave it—4—to see if they're equal. If they're not equal,
the appropriate action is to go back and add again. Thus,
another 23 is added to the total, and the counter clicks to indicate a 2. The counter is still not equal to 4, so the process is
repeated again. Eventually the counter *is* equal to the value
you are multiplying by and the appropriate action is to stop
adding and display the result (92) on the screen.

These rules for multiplying could be summarized as follows:

MULTIPLY PROGRAM

Line 1   Set the counter and the result to 0.
Line 2   Add the first number to the result.
Line 3   Add 1 to the counter.
Line 4   If the counter is equal to the second number,
display the result on the screen and stop. If
not, go back to line 2 and continue.

Notice that in our example the first number was 23; the
second number was 4. But the real power of the computer is
that this program will work with any two numbers.

Although it's hard to believe, all other calculating functions, such as division and square root, are performed by following add-and-compare programs. It is this ability to use
adding and comparing as building blocks that makes the
computer so versatile. In fact, by combining these two processes a computer can be taught to imitate, or emulate, just
about any other machine. And that makes a computer the
uniquely valuable tool that it is.

The Computer Has a Brain

The place where the computer actually does the comparing
and adding is called the **CPU** (pronounced see pee you). This
is the "brain" of the computer. All computers, no matter
whether they cost $1 million dollars or only $50, have this
"brain." CPU stands for **central processing unit**, but no one

every says central processing unit (CPU sounds so much more mysterious). Sometimes people simply refer to it as the **processor.** Or you might hear the term **microprocessor,** since the processor is extremely small. But the term central processing unit really does define the purpose of the "brain." It processes information, just as a food processor processes food. It is this device that adds and compares. Even your pocket calculator has a CPU.

The CPU can only do one thing at a time. It adds, or it compares. So it is the speed with which it does this that makes the computer such a powerful and important tool. When you use your calculator to multiply two numbers, you don't notice the time that all the intermediate steps take. Each operation is done so fast, you don't have to wait for the result. In fact, the time that the computer takes to add two numbers is incredibly small—it is the tiniest fraction of a second. There is a special phrase for the time that the computer takes to perform an operation (such as adding two numbers)—it's called the **cycle time.**

Since a computer with a faster cycle time will perform its tasks more quickly, this is one factor that determines the power of a computer. When you are buying a computer, ask to see the technical specifications of the computer. You will see the cycle time listed in units called megaHertz, or MHz. Since this is a fairly meaningless measurement in terms of figuring out how long it will take a given computer to get a specified task done, it makes sense to use this measurement for comparison purposes only. I think of this rating in the same way that I think of the miles per gallon ratings on new cars: They really don't tell you what to expect of the gas consumption of a particular car. But if the rating is higher for one car than for another, you know something about the relative merits of the two. In the same way, a computer with an 8 MHz cycle time will get jobs done faster than one with a 4 MHz cycle time, although probably not twice as fast.[3]

All computers have a CPU. In fact, the CPU is the essence

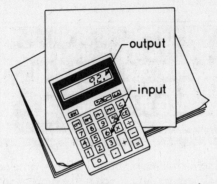

FIGURE 1-3. Input devices get information *into* the CPU. Output devices get information *out* to you. In a pocket calculator, the keys are the input devices and the screen in the output device.

of a computer. But a CPU is tiny—about the size of your fingernail. So, then, what's all the other stuff you see when you look at a computer?

Conversing with Your Computer

Think about your pocket calculator again. What use would your calculator be if it didn't have a display window? The calculator would compute the answer to your problem, but you wouldn't know what it was, since the CPU wouldn't have any way of getting its answer *out* to you. Similarly, what good would your calculator be if it had a display window and no keys? In this case, you wouldn't be able to get information *into* the CPU, telling it what to do.

All computers have **input** devices (see Figure 1-3) to get information from the outside world into the CPU. The number and function keys on your calculator are input devices. You use the keys to tell your CPU what calculations to perform.

FIGURE 1-4. The computer keyboard is really an electric typewriter keyboard, a 10-key adding machine keypad, and a few special keys which are unique to computers. The special keys, which are shaded here, are discussed in Chapter 2.

When you look at a computer you see a keyboard very much like the one on your typewriter. This is the input device that is used to get information into the computer. As Figure 1-4 shows, computer manufacturers also include a keyboard for numbers only, like the one on adding machines. This section of the keyboard is called the **keypad.**

Manufacturers include both typewriter keyboards and keypads so that you can use the one you're already familiar with. If you're a typist, you are used to numbers on the top row. If you're a bookkeeper, you are used to numbers on a 10-key adding machine.

In addition to the typewriter keys and the keypad, computers have some special-purpose keys, which we'll look at later.

There are many other input devices. If you think back to the olden days of computers, you may remember punched cards (don't fold, spindle, or mutilate!). Card readers were once a popular input device. A more modern input device is the wand that department stores use to read information from price tags into their computers. All of these devices perform the same function as the keyboard—they send information to the CPU.

All computers also have some way of getting information from the CPU out to you, such as the calculator display window. These are called **output** devices. One example is a **printer**, which is like a cross between a player piano and a typewriter. The CPU sends information to the printer, and the printer automatically types out the information. That's how you get most of your bills these days.

Another output device is the television screen. Instead of information going from the CPU to a piece of paper, the information is displayed on the television screen. But this is still an output device, since its function is to take information from the CPU and bring it to the outside world.

Computer people don't call their output device a TV, that would be too friendly. Instead, they call it a **CRT** or **VDT**. CRT stands for **cathode ray tube**, which is a fancy name for a picture tube. It is what you look at when you watch television. VDT stands for **video display terminal**, which is the same thing as a CRT.

Inside the CRT is a "gun" that "shoots" electrons out to a screen. The electrons hit a phosphorescent coating on the inside of the picture tube. When the phosphorus on the screen gets "hit," it glows. This "paints" the picture that you see when you look at television.

What isn't talked about much is that some of the electrons don't stop at the screen. As Figure 1-5 (see next page) illustrates, they continue on, through the screen, into you. In effect, you are being exposed to low-level radiation. The long-term effects of this are still unknown.

As mothers, we are always telling our children, "Don't sit so close to the TV." But as workers, we will be finding ourselves staring at a similar TV and at a distance of only 12 to 18 inches. What is clear is that more research is needed into the long-term effects of such low-level radiation.

Right now the leader in such research is the Canadian government, which responded to reports that pregnant data-entry operators working at CRTs at a Canadian hospital had an extremely high level of miscarriage or infants with birth defects. Unfortunately, it takes many years to

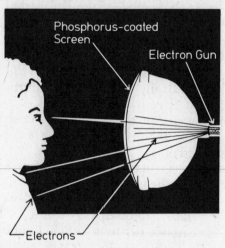

FIGURE 1-5. Some electrons don't stop at the screen, they keep going, into you. This is called low-level radiation.

conduct the long-term studies required to analyze these effects. In the meantime, pregnant women should be aware of the research and avoid working at CRTs for long periods of time.

Many home computers such as the Commodore and Atari use regular television screens as output devices. However, most business computers use better quality screens that provide sharper images. If you look closely at a television (or a computer display), you will notice that the picture is made up of dots in the same way that a newspaper picture is made up of dots. The more dots (computer people call the dots **pixels**) there are per square inch, the clearer the picture. CRTs have more pixels per square inch than a normal television screen and therefore display text and numbers much more sharply. This degree of clarity referred to as **resolution.**

Text displayed on regular TVs looks fuzzy and somewhat distorted. If you had to work at this kind of screen all day, you'd get eye strain. By contrast, CRTs used for graphics have the sharpest picture. For this reason they are referred to as **high-resolution** screens.

Although input and output devices are separate parts of a computer system, they are sometimes packaged together into a single piece of equipment, called a **terminal**. A terminal usually combines a keyboard (an input device) and a CRT (an output device). The value of a terminal is that it can be connected to a remote CPU (brain). For example, airline reservation agents work at terminals in many locations, all of which are connected to one central CPU that actually processes the ticket sales.

Some terminals have a keyboard and screen in one piece of molded plastic; others have them as separate pieces of hard-

Is this a computer?

FIGURE 1-6. Are these computers or terminals? Usually, you have no way of knowing whether or not a CPU is inside the plastic case unless you ask the salesperson. If there's no CPU inside, it's a terminal.

ware (equipment). There is a wide variety of packaging styles in the computer industry. As you see in Figure 1-6, you cannot tell by looking at a piece of equipment whether it's a terminal or a computer. Is there a CPU in the box? If so, it's a computer; otherwise it's a terminal.

Sometimes people erroneously refer to their terminal as a CRT. But, to be accurate, a CRT does not include a keyboard. The CRT is only the screen, which, when it stands alone, is called **monitor.** You often see monitors at airports showing flight arrival and departure information.

There are certain parts of a computer system that perform both input and output functions. To understand them, look at your cassette tape recorder again.

Your cassette player can be hooked up to your computer to record and play back information. For example, you can type in a mailing list on the computer's keyboard. The CPU could then send everything *out* to the tape to save it. Your tape player acts as an *output* device when it records (writes) your data on the tape. Afterward, you could send the cassette recording of your mailing list to your branch office. There the tape could be put on a tape player, the information could be played *into* the branch office computer's CPU, and the mailing list could be printed out. At the branch office, the tape player is working as an input device as it reads (plays) your data *into* the CPU.

Because the tape player has the ability to record as well as to play back electronic information, it is called an **input/output** device. Actually, computer people never say input/output device, they say **I/O** (eye oh) device, so it sounds very technical.

When you buy a computer, you get a CPU (the brain). Then you get to specify your choice of input and output devices. Figure 1-7 shows some of these devices. If you run a supermarket, for example, you might want an optical scanner as an input device to read the bar codes off packaged foods. The supermarket computer not only has an input device reading in the item you are purchasing, it also has an output device—the beeper. The beep that you hear is really a shorthand way of the CPU signaling that it got this bar code okay or it didn't get that product code and needs it again. Although the supermarket computer system seems very different from an office computer it's really only a difference between the input and output devices that are being used.

What's a Brain without Memory?

Remember our mailing list? It was recorded on cassette tape the same way a concert performance is. In both the music

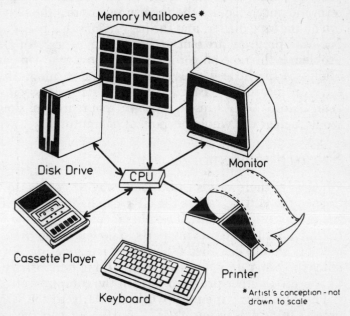

FIGURE 1-7. A computer always has a CPU (the brain). In addition, there are a variety of optional components that can be plugged into the computers. These are either input devices that send information into the CPU, output devices that bring information out to you, or input/output (I/O) devices, which are a combination of the two.

and computer examples, the value of the tape was that it acted as a storage medium. That is, you got to save the concert or the mailing list so you could replay it later.

We are going to see that the most important role the tape plays is that it acts like a computer's long-term memory, preserving the information that you've been typing (entering into) the computer. This is necessary, because when you turn a computer off, it forgets everything it knew. Without the ability to save the information, your computer would be extremely limited.

There's a problem with the computer system as I've described it thus far. Think about this for a minute. You have a program, such as Pac-Man, stored on a cassette tape (long-term memory). To use the program you press the play button on the tape player, and the instructions that make up the Pac-Man program are sent to the CPU to be performed. The problem is the tape player is slow—so slow that you can actually see the little wheels in the cassette turning. So, you have an incredibly fast CPU waiting for what seems like years before the next instruction gets sent to it. This slowness would defeat the whole purpose of computing!

Short-Term Memory

Obviously, there needs to be some way to get the instructions to the CPU as fast as the CPU can handle them. There is a mechanism for that. It's called **memory**. Like the CPU, this memory has no moving parts to slow it down. Actually, memory should be called "short-term memory" since it only holds the program you are currently using temporarily, until you use another program or turn the computer off.

The computer's memory works similarly to the way we process information ourselves. Think about cooking again. When I'm using an unfamiliar recipe, I open the cookbook and put it on the counter. I read an instruction (maybe "blend in the sugar") and I do it. Then, I walk over to the cookbook again, read the next instruction, and follow it. This is really a slow process. But when I know a recipe by heart (ie., I've memorized it) I just whiz through it—I don't need to keep walking over to the cookbook to read the instructions or look up the amount of each ingredient. I just perform each step in turn, quickly and easily. This is what happens inside the computer when the CPU is able to follow a program rapidly that is in the computer's memory.

You'll hear memory referred to almost any time there is a discussion of computers, because the size of memory determines how large a program the computer can use. This is because the whole program must be able to fit into memory at once. You can think of memory as a bunch of pigeon-hole-type mailboxes you see in college dorms or hotels. Each mailbox has the ability to hold one character of information, called a byte. If you know how many "mailboxes" your computer has, you will know how long a program (recipe) your computer can use. For example, Figure 1-8 (see next page) shows memory holding our Multiply Program.

Think of it this way. Suppose you keep your recipes on 3×5 cards and your German chocolate cake recipe takes up four cards. You have a friend to dinner and he loves the dessert and wants a copy of the recipe. If you only had one blank 3×5 card, you couldn't give him a copy of it. (One card wouldn't have enough space to hold all the instructions.) In the same way, if there aren't enough mailboxes to hold all the instructions about how to play Pac-Man, your computer won't ever be able to play that game. For this reason, all software packages have a description telling you how much computer memory (how many mailboxes) the program requires.

When you buy a computer, you always get some quantity of memory as standard equipment. Additional memory is an option, because you can usually pay extra and buy more mailboxes. If your computer comes with 32,000 mailboxes as standard equipment, for example, you could decide to buy an additional 32,000 mailboxes—thus giving your computer system a total of 64,000 mailboxes. If you'd decided to buy an additional 96,000 mailboxes, you would have ended up with a total of 128,000 mailboxes.

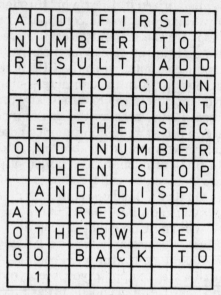

A	D	D		F	I	R	S	T	
N	U	M	B	E	R		T	O	
R	E	S	U	L	T		A	D	D
	1		T	O		C	O	U	N
T		I	F		C	O	U	N	T
	=		T	H	E		S	E	C
O	N	D		N	U	M	B	E	R
	T	H	E	N		S	T	O	P
	A	N	D		D	I	S	P	L
A	Y		R	E	S	U	L	T	
O	T	H	E	R	W	I	S	E	
G	O		B	A	C	K		T	O
	1								

FIGURE 1-8. A computer's memory behaves like hotel mailboxes. Each memory mailbox holds one character. There must be enough mailboxes to hold the entire program. In this illustration there is enough memory to hold our Multiply Program.

Computer people have their own way of describing how much memory there is. You've heard of kilowatts for thousands of watts and kilograms for thousands of grams. In the same way, thousands of mailboxes are referred to in units of kilos or **K**. Thus, 64,000 memory mailboxes are **64K**.[4] Since each memory location has the ability to hold one character, and since the "fancy" name for a character is a byte, we can say that 64K memory can hold **64 kilobytes**. So, to be really esoteric you may see memory capacity described as **64KB** which is 64,000 bytes of memory.

When you buy a computer you are also buying built-in limitations, because there is a maximum amount of memory that each computer was designed to hold. For example, a particular computer might have a maximum limit to its memory of 64K (64,000 mailboxes). Suppose this computer came equipped with 64K memory as standard equipment. Since this computer has no way to add additional memory, we say there's no way to **upgrade** this computer. If you bought that manufacturer's computer, you would be limited to programs that fit into 64,000 memory mailboxes. 64K memory is sufficient to hold many business applications programs. However, some of the newer software, such as the popular Lotus 1-2-3, require more memory. If you are buying a computer to use Lotus 1-2-3, then your choice of computer would be limited to those with larger amounts of memory. The important thing to remember is that *the order of purchasing your computer system should be software first, then hardware.* We'll come back to this concept again and again.

Memory mailboxes work just the way hotel mailboxes do. That is, each mailbox has an address. A particular program might refer to a particular address, as in "Go to mailbox 835." When a particular mailbox is referenced, the CPU doesn't have to look at each mailbox in order; that is, it doesn't look in mailbox 1, then mailbox 2, then mailbox 3, then mailbox 4, and so on. Instead, the CPU just looks in the mailbox specified (in this example, mailbox 835). Of course, this is the same way a room clerk uses hotel mailboxes. If you ask the room clerk, "Is there any mail for Room 835?" the room clerk doesn't look in every mailbox. Most likely he or she looks at the eighth row down, where the 800s are, and then in the middle of the row, where the room numbers that start with the 30s are.

This way of accessing information is called either **random access** or **direct access.** And memory is not really called mailboxes but instead (using acronyms, of course) is known as **RAM,** short for **random access memory.** (Actually, **direct access memory** would be more descriptive, but the acronym DAM was probably considered vulgar.)

This ability to go *directly* to a particular memory location adds greatly to the speed at which a program can be executed. To see why, let's look at a familiar example of direct access—using an index in a book. Suppose you are cooking a fruit pie. First you follow the directions to make the filling. Then the recipe may say, "Use any single crust recipe to make the shell." So, you look up "pie crust" in the index. The index shows that it is on page 835. All you need to do is turn to page 835, and voilà, you have the crust formula. It would be extremely laborious and time consuming to use a cookbook that didn't have an index and numbered pages. If there were no index, you would have had to flip every page, starting at the first page, and going through 835 pages before you finally found the single crust recipe you were looking for.

If there were no index and no page numbering, you would be experiencing **sequential access**. We deal with sequential access when we use a cassette tape recorder, because there is no way to "jump" into a spot in the middle of the tape. We can call the tape recorder a sequential device, because we always have to go *in sequence*, using fast forward to bypass the songs we don't want to hear in order to get to the song that we do want. Sequential access is obviously more time consuming; for this reason, tape is never used as internal memory.

In a computer's memory, direct access has the advantage of letting the CPU go directly to a particular memory location. Different segments of a program are located at different memory locations. For example, think of automated tellers. You have the choice of making a deposit or making a withdrawal. The instructions for processing the deposit begin at one particular memory location, whereas the instructions for processing the withdrawal begin at a different memory location. When you use an automated teller, one of the first instructions in the program is to compare which type of transaction you have selected. That instruction might go something like this: If transaction code = 1, then go to memory location 10,234 (which is where the deposit program begins). If transaction code = 2, then go to memory location 18,412 (where the withdrawal instructions begin).

You can clearly see that the ability of the CPU to jump to a particular mailbox, by address, makes everything go much faster, just as having an index makes it much faster to locate something in a book.

Disks Are for Long-Term Memory

Remember, the purpose for having computer memory, RAM, is to be able to hold the computer program someplace where the CPU can rapidly follow it—while you are using that particular program. But before the program can be used, it has to be played into memory from the tape or disk (more about disks shortly) on which it is stored. Computer people call this process **loading** a program into memory.

Storage

The tape or disk functions as a **storage device** (also referred to as **mass storage**). This is similar to a cookbook, which stores recipes until you're ready to open the book and use them. Computer storage devices not only hold your programs, they also hold your data, such as the names and addresses on your mailing list. And just as with a program, when this information is to be used it is first sent from the disk or tape into memory (RAM).

There's one problem about using cassette tapes for storage—they're slow. To load a program from cassette tape to memory (RAM) may take 20–30 minutes. A faster solution for storing programs is to record them onto **disks.** You can think of a disk as a phonograph record that is enclosed in a square envelope. You might think that the device that plays the disks would be called a "disk player," but computer people call it a **disk drive.** Figure 1-9 (see next page) shows a disk being inserted into a disk drive.

Disks perform the same functions as tapes, but they are much faster. Unlike cassette players, disk drives were specifically made for computers and were designed to transfer information as fast as possible. Instead of turning slowly like

FIGURE 1-9. The device that plays disks is called a disk drive.

cassette players, disk drives turn the disks quite rapidly—about 6 revolutions per second. Because of this it only takes a few seconds to load a program into memory from a disk, unless the program requires a great deal of memory. This type of storage is very fast, but it's also more expensive. Disk drives start at about $300.

A disk drive is a cross between a record player and a cassette recorder. With a record player, if you want to hear the third song on a record you can just pick up the needle and put it down on the third cut. Thus, a record player is a direct access device, allowing you to go right to the desired location without having to start the record at the beginning and wait for it to get to the song you want.

The disk drive works the same way. Instead of putting a record onto a stereo, you put a disk into a disk drive. The disk is round, but you may not realize this since it is permanently encased in a square plastic protective jacket. The disk drive has a spindle that turns the disk, much like a phonograph (except that the disk spins inside of its jacket).

Information is recorded on disks in **tracks**. This is also similar to records, except the tracks on disks are concentric circles, not a long continuous spiral like on a phonograph album. (See Figure 1-10.) As with a stereo, the "needle" of the disk drive can simply be picked up and moved to a new track. If you look closely at the disk jacket, you can see an oblong slot where the "needle" can touch the spinning disk surface. Since information can be retrieved from a particular track without having to go past all the previously recorded information, the disk drive is classified as a direct access device.

A disk drive is also like a cassette recorder. Disks are made

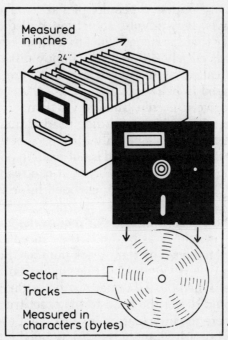

FIGURE 1-10. A disk is a computer's "filing cabinet." The capacity of a disk is measured in terms of the characters that can be recorded on it. This depends upon the recording method used by a particular brand of disk drive. Characters are recorded on concentric tracks that are similar to the grooves of a phonograph record.

out of the same media as the tapes—mylar coated with iron oxide. That's why the disk is protected by a jacket —so you don't touch the sensitive recording surface. The jacket serves the same purpose as the plastic cassette holder on a cassette tape. It keeps your fingers from coming in contact with the recording surface where the oil from your skin will ruin it. The oval-shaped opening in the disk jacket where the recording surface is exposed is a dangerous area. If you touch the disk there you may wreck its contents and be unable to "play back" your previously recorded information.

Because these disks are pliable, they're often referred to as **floppy disks** or **floppies.** Since floppies are smaller than old-fashioned disks, they are also sometimes called **diskettes** (in the same way that kitchenette means a small kitchen). Both of these terms clearly differentiate floppy disks from other larger disks called **hard disks.** Hard disks are rigid metal platters encased in heavy plastic.

The disk drive's "needle" that moves to a particular track isn't really a needle—it's a recording head similar to the ones used in tape recorders. The only difference is that the disk drive's head is able to move and is not fixed in position. Like the recording head on a tape player, it can read (play) and write (record) the magnetically encoded information that's on the disk.

To enable the recording head or **read/write head** to get to information as quickly as possible, the concentric tracks on the disk are referenced by their track number. For example, it you had a disk filled with airline flight information and wanted to see when the Albuquerque to Seattle flight was, a program would direct the head to the exact track (let's say, track 30) where the information is stored. But even this process is made faster by having the disk divided into segments called **sectors,** so that the read/write head does not have to take the time to read the entire concentric circle of the track to get to just that place where the data it's seeking is located. By being able to identify exactly which part of the track (which sector) the information is on, the head can zip exactly to that spot very quickly.

Disks come in various sizes. The two most common are 5 1/4 inches and 8 inches in diameter. The newest size (used by Apple's Macintosh and Hewlett-Packard's HP 150) is 3 1/2 inches. But, amazingly, there's no way to tell what the capacity of a disk is just by looking at. It depends upon the way information was recorded on it, not on the disk's physical size.

DISK FORMAT

The history of the computer industry has been one of fierce competition, and because of that there are not many standards or agreed-upon conventions. This is particularly apparent when you look at disk drive technology. Each disk drive manufacturer has developed a different scheme for dividing up disks into tracks and sectors. One common scheme for addressing locations on a disk is to have 40 tracks that are divided into 8 sectors, but there are many others. The technical name for the way in which the tracks and sectors are laid out on a disk is **format**.

I like to think of the tracks and sectors as parking spaces for characters. In an automobile parking lot, the total number of cars that will fit in the lot depends on how the lines are painted on. If the lines are painted far apart and there are very large parking spaces, you will obviously be unable to fit as many cars as if you marked your spaces closer together (say, primarily for compact cars).

Similarly, disk drive manufacturers use different formats (methods) for marking tracks and sectors. Depending upon which format (method) is used, more or less information (bytes) can be recorded (parked) on a disk.[5]

Disk drives are rated as either **single density** or **double density**. As you might guess, a double density disk drive can record about twice as much on a disk as a single density disk drive. Some disk drives (more expensive ones) have two recording heads—one for each surface of the disk. The effect of **double-sided** disk drives is to give you twice the recording capacity on a disk than is possible with a **single-sided** disk drive.

Saying that "the system includes two disk drives" really isn't very informative. The correct way to evaluate a computer's disk drive is by its capacity—the amount of information it can hold on a disk. This is important because the disks you use will hold your data and act as your filing cabinet. It's even more important because of the one big difference between your disk storage and a real filing cabinet. Most software requires that an *entire* data file fit on one disk. This is exactly opposite from what you used to when you use real filing cabinets. For example, when you use a real filing cabinet, to hold your client file you might have one drawer marked A–M and another drawer marked N–Z, meaning that you had so many clients that you needed two drawers to hold them all. But, when you use an electronic filing system, all the clients, A–Z, have to fit on one disk. If you have many clients you will need a computer system with larger capacity disk drives. (And if you have too many clients to fit on the largest capacity floppy disk, you'll need to use a hard disk.)

Long-Term Memory (Disks) and Short-Term Memory (RAM) Work Together

Most computer systems have a lot more disk (mass storage) capacity than RAM (memory) capacity. For example, a Radio Shack TRS80 Model 12 computer has 64K RAM. Its disk drives each hold 1.2M bytes (M stands for million or megabytes) of storage. This huge difference in capacity is necessary because memory (RAM) only has to hold the program and *some* of the data. But a disk holds your *entire* data file.[6] Usually, your data will be much larger than the size of the program that you are using to process it.

To understand the difference between memory capacity and disk capacity, let's compare typing on your typewriter to typing using a computer (in other words, **word process-ing**). A page of typewritten text is about 3K characters in length (60 characters per line times 50 lines equals 3,000, or 3K, characters). Although you might have a 120K (40-page) manuscript stored on your disk, a computer with 64K of memory is perfectly adequate. This is because as you are working, most of your manuscript remains on the disk; only what you are actually writing or editing is sent into RAM. Thus, RAM only needs to be big enough to hold the word processing program and the one page on which you are working.[7]

This is just the way you work when you type on a typewriter. While you are typing, only one page is in your typewriter at a time. After you finish a page, you put it on a stack with your other completed pages and put a blank sheet of paper in the typewriter. In a computer, the stack of typed pages is stored on a disk (Figure 1-11 on next page). The page in the typewriter is the same as the one in the computer's memory (RAM).

To be sure you see the difference between disk capacity and memory capacity, here's another example using a 3×5 card filing system. Suppose you want to send renewal notices to everyone whose magazine subscriptions have run out in January. If you were doing it manually, you would look through your card file box, picking up one subscription card at a time. You'd look at the first card and see if its renewal date was January. If it had a different renewal date you'd put it back in the file box; if it had a January renewal date you'd type up a renewal notice and then return the card to the box. You would continue this process until all the cards had been reviewed. Notice that at any one time, you would be looking at only one card while the others remained in storage.

In the same way, a computer-based filing system would only need to have one subscription card in memory at a time; all the other cards would be on a disk. On the computer-based system, the program would go out to the disk and play the information on the first electronic card into memory. Then the program would compare the renewal date to

In a computer this page is in RAM

These pages are stored on disk

FIGURE 1-11. Internal memory (RAM) only needs to be large enough to hold the part of the data file that is being acted on. The rest of the data file, like the rest of the manuscript, is stored elsewhere (on disk). Just as with a typewriter, only one page is being written or edited at a time. That page is automatically sent into RAM and replaced when another is to be used.

January. If the renewal date on the electronic card in memory wasn't equal to January, it would go to the disk and play the next electronic card into memory. It would do this over and over again. Each time, the same memory (mailbox) locations that held the first person's renewal card would be reused. When the program found a match, that is, an electronic card with a January renewal date, it would send the person's name and address to the printer. Then it would go back to the disk and play the next person's information into memory.

As you can see, the memory mailboxes work just the same as hotel mailboxes—they are reusable. When I stay in room 521, that's my mailbox. But, when I check out of the hotel, mailbox 521 will be used for the next guest.

In both the word processing example and the electronic filing example, memory (RAM) didn't hold all the data at once. It held only one page of text or one subscription card at a time. But the disk held the entire data file—perhaps 400 pages of text or 3,000 subscription renewal cards.

How big a capacity disk drive will you need? That depends on your particular type of work. Do you have 50 clients, 500 clients, or 5,000 clinets? How much information do you need to keep for each client? Do you have name, address, phone number, and payment records? Or do you have pages and pages of medical history? Do you put out a 4-page newsletter or 250-page management reports? Depending upon your situation, you will need different capacity disk drives. Chapter 2 will show how to calculate just what your disk capacity requirements are.

Storing Characters in Memory

How does a character of information actually get stored in a mailbox? To answer this question let's see exactly how the mailboxes work. That will help clear up some of the mystery about how the mailboxes get reused. Then you'll be able to really understand memory and why everyone is so excited about the silicon chips that you've heard so much about. We'll also look at how information gets moved from place to place—from the keyboard to the CPU, from the CPU to RAM, from RAM to the CPU, from the CPU to the CRT, and so on.

Computers work according to a basic property of electricity: When electricity moves around a piece of iron, the iron acts like a magnet. Look at Figure 1-12 (see next page) and think back to that familiar grade school science experiment when your teacher wrapped a piece of copper wire around a nail. When both ends of the wire were connected to a battery, the nail acted like a magnet and attracted little metal shavings from steel wool. When one end of the wire-wrapped nail was disconnected from the battery, the circuit was broken and the metal filings fell off. Without electricity the nail suddenly ceased to be a magnet. This experiment illustrated the property of **electromagnetism.** (No one knows why electricity flowing around iron causes magnetism, it is just a fact. I think it's comforting to remember that much "science" is merely describing in detail things that still can't yet be explained. For example, no one knows why gravity exists, though it can be measured and observed.)

Electromagnetism is used in many common devices. In automobile junkyards, the electromagnets that pick up demolished cars are just bigger versions of the magnetized nail. When the electrical circuit is broken, the magnetism stops

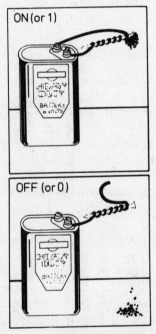

FIGURE 1-12. Electromagnetism is the basis of computing. When electricity flows around a piece of iron, the iron becomes a magnet. Here, both ends of the wire are attached to the battery, electricity flows in the wire, the iron nail becomes a magnet, and metal particles are attracted to it. When the flow of electricity is interrupted by disconnecting one end of the wire, the metal particles fall off.

and the car falls off into the junkyard pile. Electromagnetism is also used in doorbells. In an electric bell, there is an electromagnet. When you press the doorbell, the circuit is completed and the hammer gets magnetically attracted to the bell and strikes it.

Electromagnetism is also the basis for an early communications device—the telegraph. You're probably familiar with the telegraph from watching Westerns on television. Usually there is a telegraph operator wearing a green eyeshade, sitting in the telegraph office, tapping out code on a telegraph key. What actually happens is that each tap of the sender's telegraph key completes an electric circuit. The electricity flows, and an electromagnet attracts the receiver's telegraph key causing it to click.

The person who is receiving the message listens to the tapping of the key and transcribes the sounds. Perhaps two taps close together (dots) and then two taps with long pauses in between them (dashes) stand for the letter "T." Both the operators have had to agree on this representation of the letter "T" for the telegraph system to work. (In fact, Samuel Morse came up with a code for each character of the alphabet, and it is this system of codes that was named after him and called "Morse Code.")

In a similar way, a system of electrical codes is used to move information to and from the CPU of your computer. Each character of information that's entered from the keyboard gets represented by a code that for simplicity I'll call Morse code. (In fact they aren't really the identical codes, but they are similar.) When you type on the keyboard, you are sending bursts of electricity—a "Morse code"—to the CPU. Each key letter, number, or other character has a unique pattern of electricity bursts that it sends. Instead of dots and dashes, the convention for computers is to use a 0 (zero) to mean no electricity is flowing and a 1 (one) to mean electricity is flowing. (In many books the character for zero, Ø, has a slash through it to show it is not the letter "O"). An internal clock acts as a sort of metronome to keep track of the beats when no electricity flows.

Here is an example. When you type the letter "A," a particular combination of electricity bursts gets transmitted first to the CPU and then to the memory mailboxes. Suppose the pattern of electricity bursts for "A" is no electricity (0), electricity (1), no electricity for five beats of the metronome (00000), then one more burst of electricity (1). In our notation system, it looks like 01000001.[8]

Now suppose that each memory mailbox contains a row of eight wire-wrapped nails. Each burst of electricity would affect each nail in turn. Figure 1-13 shows a memory mailbox containing the representation of the letter "A." The first nail didn't get any electricity and therefore isn't magnetized. The second nail is magnetized. The third through the seventh nails aren't, and the eighth nail is magnetized.

If the CPU wanted to see if two letters were the same, it would compare them nail by nail. If all eight nails had the same pattern of magnetization, then the CPU would find the two characters equal.

In computers, there really aren't nails. In older model computers, there were iron cores, which were metal rods

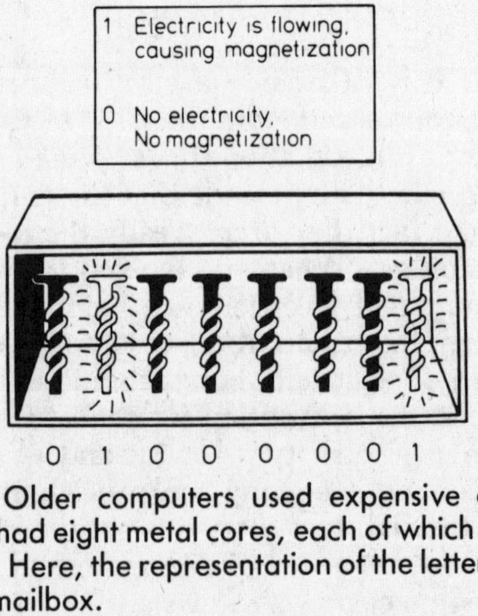

FIGURE 1-13. Older computers used expensive core memory. Each mailbox had eight metal cores, each of which could be magnetized or not. Here, the representation of the letter "A" is stored in a memory mailbox.

with wire wrapped around them; they are almost like nails but smaller—so memory was called **core memory**. This older "nail technology" had some serious problems, which limited computers from being used at home or by small businesses.

One problem with these 1960 and 1970 computers was heat buildup. The heat was caused by a side effect of electricity—friction. You experience friction when you rub things together, like rubbing your hands together on a cold day. Sometimes the effects of friction can be dramatic. For example, when you bend a wire coat hanger back and forth, it gets so hot it eventually breaks. The heat is caused by the electrons in the coat hanger rubbing against each other.

Electrons (electricity) try to move in straight lines. When electricity flows in a coil the electrons scrape against the side of the wire as they go around the turns of the coil and this friction gives off a lot of heat. Toasters have coils to generate heat intentionally, but in a computer too much heat will burn or melt the parts. Thus the computers that used core memory had to be housed in special air-conditioned rooms that had raised floors with cooling ducts underneath.

For this reason no one could afford to have a computer at home. In addition to the expense of the computer, there were huge monthly electric bills for air conditioning and running the computer. (Just like toasters or heaters, any electrical device that gives off a lot of heat uses a lot of electricity.)

Another problem was space. Space was needed for the air-conditioning system and the air ducts, as well as for memory. In fact, memory took up a massive amount of room. Think about it. Suppose you had a computer with 65,536 memory mailboxes and each mailbox had eight cores. That adds up to 524,288 wire-wrapped cores!

Newer computers—including the personal computers or microcomputers of the 1980s—use a revolutionary strategy for putting information into memory. It was discovered that **silicon atoms** could be used with the same result as metal cores wrapped with wire. Atoms of silicon are affected by electricity, too, but instead of becoming magnetized, they store tiny amounts of current as charges of static electricity.

Silicon atoms are so tiny they can't be seen without the use of a powerful microscope. This takes care of the space problem—256K memory mailboxes fit on a silicon chip the size of your fingernail. In fact, since the chips are so small they're often referred to as **microchips**, and personal computers, which use lots of them to cram a great deal of capability into little space, are called **microcomputers**.

Silicon Atoms and Characters—Bits and Bytes

Silicon is a common element naturally occurring in sand. Thus silicon is cheap; you can get lots of it at any beach. Since silicon is cheap, memory is cheap. The silicon itself is so small that it's difficult to handle alone and so it's put into a plastic "case" to make it easier to deal with. The manufacturing cost consists of the cost of purifying it and putting it on the plastic chips that are used to hold it.

I have a hard time visualizing these incredibly tiny silicon atoms, so I think of them as little creatures named Annie. (See Figure 1-14 on next page). When we send a burst of electricity to Annie, she stands up. If Annie doesn't get any electricity, she remains sitting down. If we had a microscope and could look inside a memory mailbox and see eight Annies with the first Annie sitting down, the next Annie standing up, the third through the seventh Annies sitting down, and the eigth Annie standing up, we would know that the contents of the mailbox represented the same letter "A" that we had typed in from the keyboard earlier.

Of course, computer professionals don't talk about creatures named Annie. When they talk about the things that represent

FIGURE 1-14. The computer name for a character, such as the "A," is a byte. A byte consists of eight individual silicon atoms or bits. In personal computers, a character is represented by a code that is based upon whether each of the eight bits is on or off.

characters, they call them **bits**. In the older computers, each wire-wrapped core represented 1 bit. In the newer microcomputers, each silicon atom in memory represents 1 bit.

A bit doesn't really convey any information by itself. It's merely energized or it's not; it's "on" or "off." It is the pattern of a group of bits that represents a character. As Figure 1-14 shows, the computer word for this definition is byte. Put more simply, a pattern of bits makes up 1 byte. Now that you know this you too can sound technical and talk about the number of "bits in a byte."

In our examples, we've been talking about 8 bits in a byte. Remember, there were 8 cores or 8 molecules in each memory mailbox. That's a little misleading. Most, but not

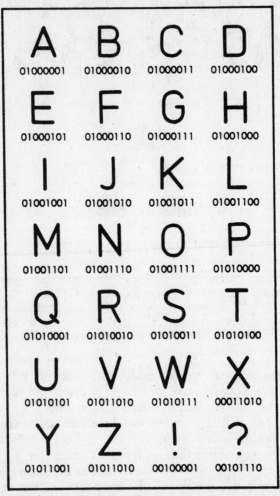

FIGURE 1-15. In computer codes, each character must have a unique representation. There must be 26 codes for the uppercase letters, 26 codes for the lowercase letters, 10 codes for the numbers 0–9, and about 30 codes for the special characters. This chart shows some of the codes of the ASCII system.

all, computer codes do have a convention that a pattern of 8 bits represents a character. But some codes, for example Control Data Corporation (CDC), use the convention that 6 bits represent a character. If you think about it, you'll see that the only requirement for designing computer codes for the English language is that there be 26 patterns for lowercase letters, 26 patterns for uppercase letters, 10 codes for the numbers 0–9, and about 30 more codes to represent special characters such as %,$#@*/? + = .

Different computers use different codes to represent characters. One popular code is called the American Standard Code for Information Interchange. Computer people call this code **ASCII** (pronounced as key). Figure 1-15 shows ASCII codes for capital letters (and some other characters).

Since each bit can be in either of *two* states—on or off—we use the prefix *bi*, which means "two," to name the code. Instead of Morse codes, they are called **binary codes**.

COMPATIBILITY

Who cares what the binary code is for any particular character? If you had a telegraph operator send a message, you really wouldn't care about the particular code that was used. What you would care about is that the sender and the receiver agree on the code they're using and that your message gets through. Similarly, you don't really care about the binary code that gets transmitted to memory when you type the letter "A" on your keyboard. All you care about is that the keyboard, the memory, and the printer all understand the same code. All that's important is that the "A" that you typed in gets printed out as an "A." If it does, then your printer is **compatible** with your computer.

To understand compatibility, think about the system of shorthand. A stenographer takes shorthand; that is, she writes some squiggly lines that represent words or phrases. A transcriber must read the shorthand and type regular English sentences from it. There are two accepted sets of rules, or codes, for shorthand—Gregg and Pitman. If the stenographer used Gregg shorthand and the transcriber only knew Pitman, then they would not be compatible, and the shorthand notes would be useless—the transcriber wouldn't be able to type from the stenographer's notes. The transcriber and stenographer must use the same system; they must be compatible in order to work together. Compatibility is discussed in greater detail in Chapter 3.

The Trend in Silicon Memory— Getting More for Less

The cost of memory has been steadily dropping. As we said, the manufacturing cost of memory reflects the work of purifying and smearing the silicon on a chip (similar to spreading peanut butter on a cracker) and then attaching wires to it to carry electricity in and out of the chip. The original chips had only about 1,000 molecules per chip, and it took eight chips wired together in order to have 1K of RAM. But in a very short time new ways to coat the silicon more densely were invented. This allowed the atoms to be packed tighter together with less space between them. (This is similar to the difference between whipped butter and stick butter. Whipped butter has lots of air between the fat molecules; stick butter is denser.) When more silicon atoms are crammed onto a chip, the price of memory drops. This is because it costs about the same amount of money to produce a chip coated with 64,000 atoms of silicon as it did to produce one with 1,000 atoms.

Memory chips (RAM chips) quickly evolved from chips with less than 1K RAM to chips holding 8K RAM each. Computers using 32K RAM chips are beginning to be produced. What this means is that the cost of memory will continue to drop. Where will it end? IBM has already produced an experimental 128K RAM chip. Wiring together eight such chips would provide over one million memory mailboxes! Although these chips are only experimental, it is estimated that within 10 years these chips will be common and cost less than $10 each. (An 8K RAM chip now sells for less than $4.)[9]

Earlier we talked about upgrading memory; that is, buying additional mailboxes for your system. What really happens when you upgrade memory is that you add additional memory chips. You add the chips by plugging them into **sockets** (spaces) on a circuit board. Figure 1-16 shows a cir-

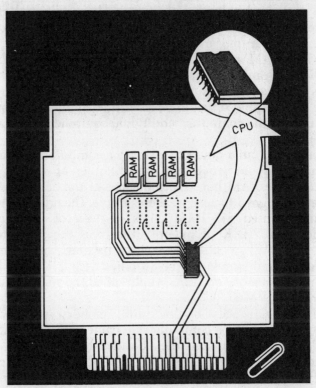

FIGURE 1-16. The maximum memory a particular computer can have is designed into the computer. This RAM board has eight sockets for RAM chips. As shown, there are 32K of RAM (four chips of 8K each). Four more RAM chips can be added to upgrade to a maximum of 64K memory.

cuit board with eight sockets. Four of them have 8K RAM chips for a total of 32K RAM. Four sockets are empty. Therefore, this computer can be upgraded by adding four more 8K RAM chips for a maximum of 64K RAM. Note that if the board had been designed with only four instead of the eight sockets shown, the computer would be limited to a maximum of 32K RAM—the amount that came standard (preinstalled) on the board, with no way to upgrade.

Many well-known brands of computers come with very little memory but with lots of space for additional chips. These extra memory chips are not included in the purchase price. For example, until recently the Apple II came with only 16K RAM as standard equipment even though almost all the programs for Apple II required at least 48K RAM. (This is a typical example of misleading advertising. The price of the Apple II would be advertised at one price, but to actually *use* the computer, additional equipment was necessary.)

The maximum amount of memory a computer can have is dependent upon two things—the number of sockets provided on the RAM board and the size of the chips that the socket is designed to accept. For example, the original Apple Macintosh used 8K RAM chips. Sixteen of these chips provided the 128K of memory that was the Macintosh's maximum. When 32K RAM chips became commercially available, Apple replaced the 8K chips with 32K chips for a maximum memory of 512K. Apple called the resulting computer the Fat Mac.

The actual process of upgrading memory is incredibly easy. If you bought a computer that came with 32K RAM as standard equipment and then decided to upgrade it, you would buy additional chips and simply push them into the empty sockets in the computer. If you have the computer store upgrade your machine, you will likely be charged an upgrade fee ranging from $30 to $60, in addition to the cost of the new chips. It's only the "mystique" surrounding computers that allows the computer stores and technicians to charge for a simple service such as plugging the chips into their sockets. Often sophisticated computer users pay for upgrading simply because they aren't knowledgeable about how their computer is built.

Circuit boards are a clever device designed to allow silicon chips to be connected together without the need for laborious wiring and soldering. Circuit boards are often called **printed circuit boards** because they use a technology of printing called resist. To understand how circuit boards are made, think about how Easter eggs are made.

In Easter egg dyeing you use a wax crayon to draw a pattern. Then you dip the egg into dye. The wax resists the dye. When you remove the wax you have a colored egg with a white pattern on it. The resist method of printing such as that used in batik is similar. A pattern is drawn on fabric with a wax resist. The cloth is then dipped into dye. Where the wax is, the dye can't get to the fabric. To remove the wax, a piece of absorbent material (such as newspaper) is put over the cloth and heat is applied using an iron. This melts the wax, which gets absorbed into the newspaper.

The "wires" on a circuit board are printed on in the same way. The plastic circuit board is dipped into molten metal, either zinc or copper, completely coating the board. The pattern of "wires" necessary to connect chips together is painted on with a resist. After the resist dries, the board is dipped into a vat of corrosive acid that eats away the exposed metal leaving only the pathways where the resist has been painted. Then the board is treated to make the resist melt and drop off, giving you fine copper or zinc "wires" where the resist had been.

These "wires" have a big advantage over conventional wires. First, they are bonded to the plastic board so they can't come loose. Since there aren't solder connections joining the wires to the board, they're not sensitive to jarring and breaking loose, making them more durable. Also, these "wires" take up less space then regular wires, thus conforming to the small space requirements of personal computers. The printing process that generates the "wires" is automated, which makes the cost much lower than if real wires had to be soldered into place by hand.

Circuit boards have a lot of advantages, but they have hazards associated with their manufacturing. The corrosive acid that eats away the unwanted metal on the board has to be washed off and disposed of. Unfortunately, this isn't always done as carefully as it should be. Some of the cancer-producing material that is used in this process has been contaminating the water supply in the Silicon Valley communities in California. This is one of the deceptive things about computer manufacturing—it looks like such a "clean" industry that you often don't notice the pollution that surrounds it.

Circuit boards aren't only found in computers. They're used anywhere that saving space is important. If you look carefully at the electronic items around your house, you'll notice that they use circuit boards. Garage door openers and color televisions, for example, have printed circuit boards inside them. Another name for a printed circuit board is a **card**. If you see a description for a **memory card** or a **RAM card**, you know that it's the circuit board with the memory mailbox silicon chips on it.

Devices that use silicon chips and printed circuit boards are called **solid-state** devices. In general, they are very durable. In fact, the only thing that can harm memory or the CPU is a **surge of electricity**. If more electricity goes to them than is expected, they can burn out.

The current that your power company supplies isn't perfect—sometimes it dips and you get less than you expect. When lights dim, it's sometimes referred to as a brownout. And sometimes it surges and you get too much current; lights then seem a little brighter than normal. As a general rule, a dip in current doesn't physically hurt the silicon, although you might lose the information that you have in memory if power drops low enough so that the molecules revert to their at-rest position. It's the power surges that you really need to protect against. For this reason, it's a good idea to plug your computer into a **surge protector** such as the one in Figure 1-17. The surge protector then gets plugged into the wall outlet. The surge protector works something like a filter—it filters out power spikes. That is, it prevents a power surge that comes from your electric line from going into your computer. For this reason, it's also known as a **line filter**. Line filters cost under $100, and you can buy them at any computer store.

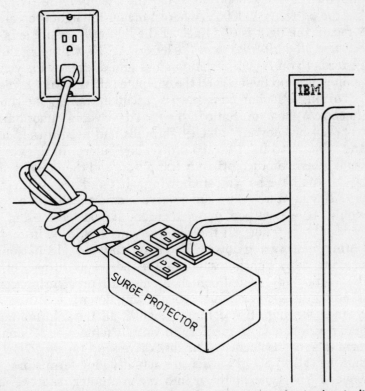

FIGURE 1-17. To prevent damaging a computer, don't plug it directly into a wall socket. Instead, plug the computer into a surge protector (line filter) that is then plugged into the wall socket. When your power company's electricity surges to levels above 120 volts, the surge protector acts as a filter and the surge does not reach the computer.

Short-Term Memory Versus Long-Term Storage

Earlier we said that memory (RAM) should really be called short-term memory. To see why, let's do a mental exercise. Visualize some shoeboxes. In each shoebox there's a row of eight lightbulbs, each of which could be turned on or off. Suppose the first shoebox looks like the one in Figure 1-18. It has the pattern 01000001, where 1 means the light is on and 0 means the light is off. Each of the lightbulbs represents a bit, and all 8 bits represent the letter "A." Now suppose that yesterday you typed in some names and addresses for your mailing list and then turned the computer off and went home.

Turning off your computer is just like unplugging your shoebox. When you come in the next day, you will not have any way of knowing which lightbulbs had been turned on and which had been turned off the day before. In the same way, when you turn off your computer, and then turn it on again, you have no way of knowing what had been in memory. If you were using older, core memory, all the cores would have lost their magnetization. If you were using a personal computer, all the silicon atoms would be in their resting state (sitting down). The RAM chip acts just like our shoebox with lightbulbs—once the power is turned off, there's no way to tell which atoms had previously been standing up and which had been sitting down.

Since turning off a computer causes all the information that was formerly in memory to vanish, some type of long-term memory is needed. The disk drives and tape recorders that we talked about earlier are actually long-term storage devices. You type material into the computer (short-term memory) and give a command to transfer the information to the storage device. You can then turn off the computer knowing that your information was saved to tape or disk for future use. Next week or next month, or at any time you could turn the computer back on, give a command to read (play) the tape or disk into memory, and thereby restore

Newer silicon-based computer

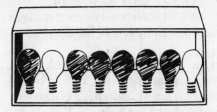

Older core memory

Imaginary shoe box computer

FIGURE 1-18. Imagine that a RAM location is a shoebox with eight lightbulbs in it. Depending upon which bulbs are lit, any character can be represented. If the power is turned off, all the bulbs will get dark and there will be no way to know which character has been stored. Similarly, when power is turned off to core memory or silicon RAM, there's no way to know what was formerly stored there.

your information. Then you could continue your work, whether it was sorting your mailing list into zip code order, editing your manuscript, or doing budget projections.

Before personal computers became the standard in office automation, memory typewriters were very popular. Actually memory typewriters *are* computers. But they have been

designed to restrict you to only doing typing. These typewriters have a CPU, some RAM (enough for a page or two),
a keyboard, a screen, and a printer, but they have no disk or
tape for long-term storage. That means if you typed something into RAM—such as a form letter—and want to store
it, *you can never turn off the typewriter*. If you did, you
would "lose" the form letter and would have to reenter it.

A point that we made earlier and is worth repeating is
that the reason for having short-term memory is that it is
very fast to access—faster than a tape or even a disk. The
reason it's so fast is that it works without any moving parts.
Instead it only uses the flow of electricity through silicon
atoms to send and receive information. The rule of thumb in
thinking about the speed of information transfer is that anything with moving parts is operating on human-scale time,
which is slow. Things such as memory, with no moving
parts and that work at the speed of electricity, are extremely
fast—so fast that as humans it is hard for us to comprehend
their speed.

Things with moving parts, such as disk drives, cassette recorders, and printers, are not only slower but they will eventually wear out or break down. Things without moving
parts don't wear out or break down. So you can see that the
CPU and memory (RAM) are not only the fastest parts of a
computer, they are also the most reliable. Therefore, these
are the parts that are the safest to buy used.

Recording Information onto Tapes and Disks

We know that information is represented in memory by a
combination of bits set in a particular pattern. We also
know that each memory location (mailbox) has the capacity
to hold one character, which in personal computers is composed of 8 bits. Depending upon the pattern of bits in a
single memory location, any letter, number, or special character can be represented. And we also know that this information gets stored as long-term memory on a tape or disk.
Now we will see exactly how the information gets stored.

Recording tape is made out of plastic (mylar) coated with iron oxide. The coating material is like a bunch of iron filings stuck onto the plastic tape with glue. These pieces of iron can be magnetized. Depending upon the patterns of iron oxide being magnetized, different tones are represented, or recorded.

The recording head in a cassette player is electrified and has the ability to either magnetize the iron oxide pieces or to sense the presence or absence of magnetization. When the head is magnetizing the iron oxide, we say the tape is *recording*. When the head is sensing the presence or absence of magnetization, we say the tape is *playing*.

The same process is used in disk drives. When referring to computers, you use the word **write** instead of record and **read** instead of play. That's why the recording head in a disk drive is called the read/write head.

You can erase a disk just like you can erase a cassette tape. When you erase, what you're really doing is magnetizing all the particles of iron oxide. Thus, the erase button is really a magnet. This means that any magnet can affect the information stored on a tape or disk. Remember the discussion about how bells work? If you lean a disk against your telephone and it rings, the magnet that makes the bell ring will ruin the information stored on that disk. So will the magnet in your pot holder or the metal detector at the airport. So be careful with your disks.

The CPU and Its Peripherals

Figure 1-19 (see next page) shows a typical computer system. Notice that the CPU is in the center and optional devices are arranged in a circle around it. Another name for the outer circle is periphery, and it is this word that is the root for the real name of the options—**peripherals**. As the word periphery implies, peripherals are of lesser importance. A computer may or may not have any of these peripherals and still be considered a computer.

All the peripherals may be directly connected to the CPU by a wire. Computer professionals call that **hard-wired**. But many times the peripherals won't be in the same physical location as the CPU and memory. Airline reservation systems typify such a system. The reservation agents have terminals (CRT and keyboard) in hundreds of cities, whereas the CPU, memory, and disks are centralized in one place. For example, Eastern airlines has its CPU and all its data stored on disk in Miami, Florida. Eastern could have strung wires to Miami from each of their reservation agents located throughout the country, but since the telephone company's wires are already in place, it is easier to just use the phone company's system.

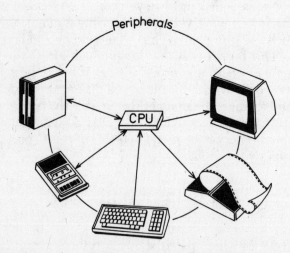

FIGURE 1-19. Optional devices are called peripherals. These devices are not essential to the operation of a computer and may be thought of as on the periphery or edge of the computer.

Modems

Telephone wires weren't designed to carry the same type of electrical signals that computers use. Since telephones are designed to carry sound, a special device has been developed, called a **modem**, that converts the computer's electrical impulses into sounds. Inside the converter is a horn that makes a high-pitched tone every time it gets an electrical impulse and a low-pitched tone when a beat passes without its getting an electrical impulse.

The modem has a cradle in which you lay the telephone receiver. The tones go into the mouthpiece of the telephone and are transmitted in the same way your voice is carried when you talk on the telephone. There is another modem at the destination that converts the sounds back into electrical impulses. These impulses are sent on to the CPU. Figure 1-20 (see next page) shows how modems help transmit data between an airlines reservation agent in one city to a centralized CPU located thousands of miles away.

To see exactly how the modem works, let's look at what happens when a reservation agent in San Francisco types an "A" at the keyboard of her terminal. First, the keyboard sends the modem the set of electrical pulses that represent the binary code for letter "A" (01000001). Next, the modem converts the eight electrical impulses into eight sounds—one low-pitched tone, one high-pitched tone, five low-pitched tones, and one high-pitched tone. These tones go into the mouthpiece of the telephone receiver and are transmitted over the telephone wires to their destination in Miami.

These tones are "heard" in the earpiece of the telephone receiver in Miami. There, the receiver is also lying in a modem. Only this modem works in reverse, transmitting an electrical impulse every time it gets a high-pitched tone. In Miami, the modem "hears" the eight sounds and converts them into no electrical impulse, one pulse, no pulses for five beats of our clock, and one electrical impulse. In this way

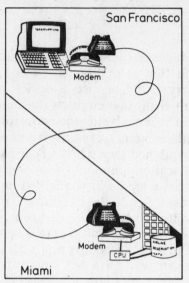

FIGURE 1-20. You can use telephone wires to connect a terminal to a remote computer. To do so you need a modem at each end.

the CPU gets the binary code for "A." To the CPU there is no difference between getting the "A" transmitted via the modems or getting the "A" directly from a terminal that is hard-wired to it.

When the computer in Miami wants to send the flight information that was requested back to the terminal in San Francisco, the process is merely reversed. Thus modems can act to both transmit and receive data, converting electrical pulses into sound (modulate) and sound back into electrical pulses (demodulate). The name "modem" actually is short for *MO*dulator/*DEM*odulator. (The word "modulate" refers to tones. For example, if you talk in a monotone, someone might tell you to modulate your voice so you sound more interesting.)

There are some special considerations about using modems. First, the telephone company ran its wires to transmit human speech, not computer-generated electrical impulses.

Since speech is relatively slow, the wires were designed to carry electrical pulses at a much slower rate than the speed at which computer wires (called **cables**) transmit it.

If you think of electricity as water, you can visualize the difference between the amount of water (electricity) you can send through a garden hose (wire) and that you can send through a large pipe (cable). In other words, computer cables carry many more pulses of electricity per second than do telephone wires. If you use the regular phone lines, called **voice-grade lines**, the terminal and CPU will be sending signals (tapping out "Morse code") at a faster rate than the telephone lines can handle, unless something is done to slow down their transmittal speed. One of the modem's functions is to regulate the speed at which the electrical pulses are sent.

The speed of the pulses of electricity in a computer is measured in units called **baud**. This word isn't an abbreviation for anything; it simply honors a Frenchman named Baudot.[10] You can buy modems that operate at different speeds (baud rates). To give you some sense of what the different speeds mean, a personal computer that's hard-wired to its terminal usually moves information at 9600 baud. But the fastest speed that voice-grade telephone lines can handle is 1200 baud. You can buy a modem that will slow computer-generated signals down to that 1200-baud speed. If you watch someone using a 1200-baud modem to communicate with a remote computer, you'll see that it takes a much longer time for information to show up on the terminal's screen than usual. This is because it is being transmitted one-eighth as slow as normal. The 1200-baud modems are fairly expensive; they cost about $300.

It's possible to buy a less expensive, but slower, 300-baud modem. When you see a terminal that's connected to a remote computer via a 300-baud modem, it is running at a speed $\frac{1}{32}$ as fast as when it's hard-wired. At that speed, it's actually quite agonizing to watch the terminal screen, since you must watch letters get transmitted one at a time to finally spell out the information you are waiting for. Although 300-baud modems are now available for under $100, the savings may not be worth it when you consider that if you're connecting to a remote CPU, your long distance phone bill will be higher since the information is coming over the line four times slower.

Once the telephone company realized that computers would be transmitting data over its lines, it began installing special lines that can carry electronic information at much greater speeds. These special high-quality lines, called **data lines**, can be used by companies, such as the airlines, that are willing to pay more money for more efficient transmission of data. Special high-speed modems operating at 9600 baud (or even faster) are used to transmit information over data lines. Naturally, using these lines is more expensive. For this reason most ordinary computer users continue to use the regular telephone lines.

Another consideration about using a modem is that the modem that you use must operate at the same speed as the modem at the other end. For that reason, many 1200-baud modems have a switch that allows you to select the speed that you will be using—either 1200 or 300 baud. To visualize why the modems at either end have to be running at the same speed, imagine trying to take dictation if someone were speaking much faster than you could write. You simply wouldn't be able to keep up.

The type of modem that we've been describing is also called an **acoustic coupler**. It's called "acoustic" because its based on sounds—those high- or low-pitched tones that the horn makes. Acoustic couplers have a drawback—they are sensitive to background sound. For example, suppose a truck were going by and honked its horn. That might get translated into an electric pulse and cause some erroneous piece of information to be transmitted. (An extra digit might be sent to an airline computer and you would have a seat reserved on flight 192 instead of 92.)

There's a funny story about erroneous sounds and a hotel chain that had a central computer for reservations and accounting. The main hotel had a large courtyard with live birds. For the longest time the staff couldn't figure out what was causing the computer's recordkeeping errors. Finally, someone realized that the squawks from the parrots were being picked up by the acoustic coupler!

To reduce problems caused by background noise, acoustic couplers use plastic, sound-deadening "cups" to put the phone receiver in. But that still isn't a perfect solution. A better solution is the type of modem that doesn't use a horn at all.

To understand how a modem can work without a horn, think about how your telephone works. Your voice is not actually transmitted down the telephone wire. Instead, when you speak into a phone the sound waves from your voice press on a diaphragm (like a membrane). The pressure exerted on the diaphragm causes electricity to flow in a pattern down the telephone wire. At the other end the electricity makes the diaphragm in the earpiece resonate, generating sound waves. (You can see this effect when you look at a stereo speaker. The speaker covering pulsates as sound is generated.)

In an acoustic coupler, the electrical impulses from the computer are turned into sounds by the modem and are then turned back into electricity by the telephone receiver for transmittal over the telephone lines. Eventually, someone realized that this was a silly waste of effort. Why go through the intermediate step of creating sound at all? Why not just have the modem convert the CPU's electrical codes directly into the type of electrical pulses transmitted by the telephone receiver? This type of modem can be plugged directly into your telephone line by using the same kind of modular plug that you usually use with your telephone. In fact, you can get a double plug and plug the modem in to your phone line permanently, without having to keep connecting and disconnecting your telephone every time you want to use the modem. Remember our discussion about printed circuit boards? Another way to purchase these new modems is in the form of a printed circuit board.

Adding Components

As illustrated in Figure 1-21, inside your computer is an area with a row of slots. The entire area is electrified, and individual circuit boards with different functions can be inserted into the slots.

That's how the computer store "tailors" a computer to your special needs and how your computer can be upgraded if you decide to add some new features at a later date. This means that another measure of value in a computer is how

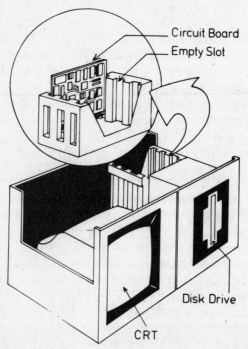

FIGURE 1-21. A computer is designed using a modular approach. Inside the computer is a box with slots. The computer store tailors your computer to your specifications by inserting the circuit boards required to handle the peripherals you have chosen.

many empty slots are there. If there are many empty slots, you have a lot of options for future expansion. If there are very few additional slots, you will be pretty much limited to the features that originally came with your computer.

There are many special-purpose boards. Each peripheral that you use in your system requires a special-purpose card to control it. For example, if you wanted to have a color monitor instead of a black-and-white display screen, you'd have to buy the color monitor as well as a special-purpose card, called a color **controller card**, or a **controller board**. Think about it: even with a black-and-white television set that's being used as a computer screen, how does this home television "know" to paint an "A" on its screen when it gets the combination of electrical pulses 01000001? It knows because there is special wiring on a controller board that translates the computer's electrical pulses into the kinds of signals that the electron gun in the television expects in order to fire the particular pattern required. You can't just plug a television into your CPU. You must also insert an appropriate controller card into one of the empty slots in the computer. This tells the television how to act like a computer screen.

Another example is the disk drive. For example, it doesn't "know" how far to move its read/write head when it gets the signal from the CPU to move to track 3. It is the wiring on the disk controller card that translates the signal to move to track 3 into a signal to move, say, ⅛ inch. The circuitry required to actually get the drive to work is located on the disk controller board. That's probably why the disk drive is called that instead of being called a disk player—it only drives, or turns, the disk, whereas its controller board is the true navigator.

Disk controller cards typically can control from one to four disk drives. So the first disk drive costs more to install than the second, third, or fourth drives, because when you buy the first drive you also have to buy the disk controller card. For example, the first drive might cost $500—$200 for the disk controller card and $300 for the disk drive itself. If you added a second disk drive it would only cost $300, since the disk controller card would already be in place.

Many people find the idea of controller cards confusing. They wonder why a disk drive or CRT manufacturer just doesn't put the circuitry that's on the controller card inside its product and eliminate the need for the card altogether. The main reason is because peripheral makers want to sell their brands of equipment for use on as many different computers as possible. Since each computer has its own unique way of electronically communicating with its peripherals, one standard disk drive would not work with all computers. Thus, the peripheral manufacturers build their standard equipment but create a different controller card for each brand of computer. In this way I could buy the same brand of color monitor for my IBM-PC as an Apple IIe owner. But we would each have to buy special CRT controller cards. I would need an IBM-PC controller card; the Apple IIe owner would need an Apple IIe controller card.

Additional controller cards can also be inserted to upgrade your screen so that it will display more characters on a line. On less expensive home computers the CRT controller card may only show 40 characters per line on the screen, instead of the 80 or more characters offered by business systems. If you wanted your system to be used for word processing, a 40-column controller wouldn't let you see how your document would really look until you printed it out. (There are between 60 and 70 characters per line on a standard page.) To remedy this problem, you can upgrade the computer by replacing the standard CRT controller with an 80-column CRT controller.

Some home computers have no way to upgrade to 80 columns. They come with a 40-column card installed and no one manufactures an 80-column card for these particular computers. If you're intending to do serious word processing on a home computer, it's best to make sure that there is an 80-column CRT card available for the brand you are buying.

Home and Personal Computers

This is a good time to start discussing the differences between a **personal computer** and a **home computer**. Typically, a home computer is less powerful and has fewer options available. That is, a home computer is designed for use at home. There's no need for 80-column word processing if you are using the word processing to write to your friends. There's no need for a high-resolution screen if it's used primarily for games and not to generate business graphics for management reports.

A personal computer is a small business computer that is personally yours. It sits on your desk at work and you don't have to share it with anyone. If you have a business that you operate out of your home, you'd still want a personal or business computer. It would not be called a home computer even if that's where it is physically kept. It's the level of technical sophistication that separates the personal computer from the home computer. We'll discuss more of the differences in Chapter 3.

Actually, there's not much you can do with a home computer. It's good for games, for teaching children how to write programs, and for simple things like balancing checkbooks (if you really need a computer to do this), but it's not very good for anything else. This is something that people in the computer industry don't talk about very much since home computers represent millions of dollars worth of business. This just shows you what the miracle of marketing can do. It seems that once owning a computer became trendy, everyone had to have one. But there are thousands and thousands of home computers collecting dust in the same hall closets that hold hoola hoops and parcheesi sets.

Computerized Information Services

Home computers can be transformed into an exciting information tool, however, by using them as a terminal and connecting them by a modem to remote data banks such as CompuServe, The Source, and Dialog.

These computerized information services are just the beginning of a future in which huge information utilities will exist. Already, the centralized computers of CompuServe have enormous disk drives that hold all kinds of pre-recorded information, such as encyclopedias, airlines schedules and fare information, and the *New York Times Index.* People with their own computers (or even their own terminals) are now able to dial up the computer utility and retrieve the information they want—be it the latest Dow Jones stock report or information to help a sixth grader do her homework.

Many of these data banks can now be reached by a local telephone number, so you don't even have to pay a long distance telephone bill. The local number connects you into the remote computer where, before you begin work, you have to identify yourself with a password. The password is yours, uniquely, and serves to maintain billing charges to your account. This password identifies you in the same way that your password or special ID number identifies you when you use an automated teller at the bank. Generally, your bill is based on the amount of time you're connected to the computer and a fixed monthly subscription charge. You may also be charged an additional fee for certain information. For example, looking up information on the Dow Jones ticker tape may entail a special surcharge.

Many of these centralized data utilities will also be maintaining libraries of computer programs that subscribers will be able to use. This means that you will be able to use an expensive and sophisticated program that your computer itself might not be able to run, simply by hooking up to a larger computer via a modem. Your computer will be acting like a terminal, except that you will use your own disks or tape to retrieve and store your data once you are finished using the remote program.

This is the future! Catalog shopping from your home, where you peruse the Sears catalog complete with pictures, order merchandise, and have your charge account automatically debited. Already the major banks, such as Citibank and Bank of America, offer a bank-by-computer service where you can transfer funds from savings to checking and pay bills just by keying in those requests from your terminal at home. It is a curious fact, however, that at the present time the banks are charging as much as $8 per month for this service, even though they benefit by having you do the work that a bank employee would otherwise have to do.

If all you wanted to do was have access to remote data banks, you wouldn't even need to buy a home computer. You could just buy a computer keyboard that connects into your television screen and a modem. The combined cost of this device might be about $150. You could then use your telephone to connect up with remote computers that can provide you with many of the capabilities of having your own computer.

Inside the Plastic Shell

Manufacturers have made varying decisions about how to package the components that make up their computers. All computers have the same basic kinds of innards, but they are packaged into their plastic shells in a wide variety of ways. Sometimes all the parts are packaged together into one solid plastic shell. But most often the parts will come in one, two, or three different pieces, connected together by external cables. The CPU, RAM, and screen might be in one piece, the disk drives in another, and the keyboard in another. Or the keyboard, CPU, RAM, and disk drives might be together and the screen separate. The packaging combinations seem endless.

The approach to computer design is similar to the way cars used to be manufactured. In the fifties when you went to buy a car, a salesperson showed you a long list of options. You decided which options you wanted: what color the car would be, the upholstery fabric, and so on. (Now, of course, the salesperson tells you that the latest shipment from Japan has come in and you can have your choice of two or three basic combinations). But, computers are still built and sold like cars of the fifties; just about everything is an option. These options are called **modules**, which are like the components of stereo systems.

The option approach to manufacturing computers is called a **modular** approach. The CPU, or brain, is the only standard equipment. All the other parts can generally be upgraded or changed, either at the time you buy your computer or later.

Since computer systems are so modular, careful shopping is required for a good price comparison. Some systems include the cost of options, such as disk drives or printers, whereas in other systems these modules cost extra. Less obvious are the hidden options such as the controller cards described previously. In some systems (such as the IBM PC), even the **ports** (outlets) that the peripherals get plugged into are extras which must be purchased.

Buying a Computer

At this point, we've discussed all the components that make up a computer system, although we will go into greater depth about some of them in Chapter 3. And you've learned a great deal about how to evaluate a computer. In fact, what you've learned is already much more than many computer salespeople know. So you see, computers really are easy—it's just that most of the books about them are hard!

Before going on, let's review some of the basics that you will now consider when buying your own computer.

First, and most important, you know that when you buy a computer you are really buying *limitations*. You are buying the ability or inability to plug in various peripherals. Does a particular computer allow you to plug in disk drives? If it does, how many disk drives can you plug in and what capacity are they? How much memory does the computer come with? Can you upgrade it by adding more memory? If so, what's the maximum amount of memory that the system permits? Is that enough memory for you? Will it allow you to use the program that you want? What kind of screen does the system come with? Will it allow you to have 80 characters across or will that cost more? Is it permanently limited to 40 characters on a line? Can you do graphics on the system? If so, is that higher resolution screen included in the purchase price or will that cost more? If you don't need graphics now, will you have the ability to upgrade the system to have that capability at a later date?

CHAPTER 2

Something about Software

You have read about what a computer is and how it works. But what can it really *do?* The true value of a computer is that it can be programmed for a variety of tasks. But what kinds of things are appropriate to use a computer for? Are there things that you can do faster and more easily without a computer?

As you might expect, computer people have a special name for the things that computers do for you—they're called **applications**. But, which applications are appropriate for a computer to work on? And do you write your own applications programs, hire someone to write them for you, or buy **application software** off the shelf?

This chapter answers these questions, and in the process, helps you make sense out of all the advertisements for application software that you see not only in technical journals, but on television and in popular magazines.

Don't Write Your Own Programs

One of the most common misconceptions about using computers is that you need to learn how to write programs in order to get the full value out of your equipment. This attitude is mainly due to the emphasis that most schools put on teaching computer programming. Most universities have computer science departments whose primary purpose is to train programmers, not to teach computer literacy. They're not interested in creating a large population of well-informed and knowledgeable computer users. In fact, their objectives are quite to the contrary. If you think I'm exaggerating, just look at the curriculum of most introductory computer classes. Most introductory computer classes teach technical concepts such as number theory—the hexadecimal and octal arithmetic on which binary codes are based—furthering the idea that computers are inaccessible and difficult. The truth is that there is absolutely no reason why anyone other than high-level computer programmers needs to know such technical information, and it certainly doesn't belong in a beginners' class. The result of this approach is to discourage people from learning how to use computers. The introductory classes are overcrowded and hard to get into, but the drop-out level is high and the more advanced classes are underenrolled. The introductory classes have served their purpose—they have kept the ranks of computer professionals small enough to maintain high salary levels.

Unfortunately one of the main effects of creating this kind of computer mystique is to discourage women from using computers. It validates the popular misconception that you have to be good at math in order to work with computers. That simply isn't true (unless your career objective is to become a hardware designer).

It's worth repeating that there's no reason to learn how to program, unless you want to pursue it as your career or take it up as a hobby (instead of knitting or crossword puzzles). The most efficient way to use a computer is to buy someone else's program. This has not always been true but has become so within the last few years with the creation of new **tailorable software**. When you think of tailoring, you think of alterations—of taking in the waist or shortening a hem to make clothes fit perfectly. In the same way, you can tailor software to adjust a program to your own specific requirements.

Let's see exactly what tailoring a program means. In the

olden days, six or eight years ago, if you bought a program, you could not alter the way it had been designed to deal with the data. For example, if you bought a mailing list program, it might have been set up to print address labels consisting of three lines. If you wanted to use the program to keep track of people at their workplaces, and have a label with the person's name on the first line, the company's name on the second line, the street address on the third line, and the city, state, and zip code on the fourth line, you couldn't use this software because it did not allow for four lines on the address label. Also, the program might have been written to accept only five-digit zip codes, in which case you wouldn't be able to use it for Canadian addresses, since Canada uses a six-character postal code.

By contrast, the new tailorable software is set up so that you, the user, can specify the characteristics of your data. You can determine the number of items you will include, such as the number of lines on your address label, as well as the size of each of these items, such as five or six digits in the zip code. In this way people with a wide variety of needs can use the same applications program simply by tailoring it to their own specifications.

As you might imagine, tailorable software is a little more time consuming to set up, or **install**, initially because you have to specify in some detail the type of data you will be using. That is, for your mailing list you would have to define the number of characters you will allow for each name, the maximum number of digits in the zip code, and so on. But that chore is only done once, and it's a lot less work than writing a special-purpose program yourself.

Problems with Programming

Although the existence of tailorable software means that you don't have to create (write) your own programs, many people still think they should. In fact, it's not only unwise to program the solution yourself (or even to hire a programmer to do it for you), it's downright dangerous. There are horrible problems that you can create for yourself when you write even the simplest of programs (or have them written for you). Let's see why.

One is that the time and effort required to write a program is considerable. This is illustrated by the fact that most software companies that sell applications programs measure the time it takes to write a program in units called **man-years** (no, they haven't yet started to call them person-years). Obviously, it's much better to spend $350 to buy a 10 man-year program (five people working 2 years) than it is to spend 10 years writing the program yourself.

Another drawback to having custom programs written is that it's hard to estimate how long it will take to write the program. (This is something that most programmers have trouble admitting. When I give estimates of how long it will take to get a new program working, I usually work out my best estimate and then multiply by 4, and, if anything, it ends up underestimated!) Oh, the program will *almost* work, but with software *almost* is not good enough. Imagine how you'd feel if a bank used a program that almost worked to keep track of your bank balance. Maybe it worked, except when you deposited money on the same day that you wrote a check, in which case the deposit showed up as a withdrawal.

It's very time consuming to get a program to work perfectly. To see why, keep in mind that a program is just like a recipe, and you run into the same type of problems when you try to write a new program as you do when you try to create a new recipe.

Suppose you are creating a new type of cake and you just tested your recipe. You may have found that it wasn't sweet enough. To solve that problem you decide to add more sugar, but that might create a new problem—it won't rise properly. So now you add more baking soda to the recipe. But it still may not taste quite right, and you figure out that the problem is not enough salt. But when you add more salt to the recipe, it tastes sweeter. So you have to go back and readjust the amount of sugar. And don't forget that each time you readjust your cake recipe, you have to bake another cake to see how it turns out.

The point is that it gets time consuming, and therefore expensive, to create a new recipe from scratch. It's the same thing with programs. After a program is written, it has to be tried out with real data (ingredients) before you can be sure it works. In this way you find the mistakes, called **bugs**, and correct them, which is called **debugging**. It's all very unpredictable—you never know how long it will take before *all* the bugs have been corrected. Remember a program isn't usable until it works 100 percent of the time. The program may be 98 percent complete for months, but until the last bugs are out, it simply isn't usable.

There are lots of programmer horror stories about all the weird and unpredictable things that happen when an unforseen set of circumstances occur. (After all, the computer just blindly follows preset instructions and cannot question them.) For example, when automated tellers were first introduced in San Francisco, the program assumed that when you lifted the drawer to get your money, that was the end of the transaction. One day someone tried lifting the drawer a second time after already having received cash from a withdrawal. The programmer who had written the automated teller program never anticipated anyone doing something like that. So the customer got another bundle of cash. It was the same amount as in the original transaction, but this time the amount wasn't deducted from his account. This worked so well, he did it again and a fourth time as well. Luckily for the bank, the inventive bank customer was a programmer himself and reported the incident to them.

Lots of programs go crazy on leap year, because the programmers who wrote the instructions for calendar progressions forget to provide for a twenty-ninth day in February every fourth year. When February 29 rolls around, the program simply can't deal with it.

The point of these stories is that a program may appear to

work, but errors may not show up for months, or even years, until some unique set of circumstances occurs.

No matter how hard a programmer tries to simulate real-life situations, there really isn't any way to create truly accurate test data except by using the program in real life. Even then unique circumstances pop up from time to time that cause the computer to go **down**. (Of course "down" just means broken. But, heaven forbid that we actually say the computer's broken—then the computer sounds like an ordinary car or a toaster, and not at all like a mythic brain.) When a computer is down it may be physically broken—maybe a moving part on a disk drive isn't working. But often the computer program has encountered a bug while it was processing data. When the program has no instructions about how to respond to an unforeseen situation, it may simply stop or, more infuriatingly, repeat the same question over and over again. For these reasons it is better to buy prepackaged software whenever possible. The better-known brands of software have been tested for years in real office situations, and all the bugs have been identified and removed.

Before Tailorable Software

Back in the "old days," before tailorable software, there may not have been an off-the-shelf program that met your particular data needs. You then had several choices. You could buy a software program that was almost right for you and hire a programmer to adapt it to your exact needs. But often such adaptation wasn't even possible, because most prepackaged software was designed to prevent programmers from getting into the instructions and changing them. You could also hire a programmer to write a program for your specific needs. Or you could change the way you had been doing things to accommodate the existing software. As the following example illustrates, none of these solutions is very good.

If you owned a dress store, you might have wanted to purchase an inventory program that kept track of what garments were in stock, deducted the quantity sold each day from the amount in inventory, and generated a reorder report each week. But such a prepackaged program might have had built-in constraints such as the maximum number of digits it allowed for the identification number assigned to each dress in stock. Perhaps the maximum number was seven, but you were using a nine-digit ID number. You had a difficult choice to make: You could redo the way you had been keeping your inventory to accommodate the computer system, hire a programmer to alter the packaged software, or if that wasn't possible (and it generally wasn't), hire a programmer to write a new program just for your store. If the clothing distributors that you ordered from required that you use their nine-digit identification code, you had no choice—you needed a new program.

Managers often chose not to have a custom program written and instead altered the old (manual) way of doing things so that procedures would agree with what the packaged software expected. This decision was understandable given the uncertainty and expense of custom programming, but it often was the wrong thing to do too. After all, if the computer was supposed to serve you, why should you have to change procedures that worked well just to suit the computer program?

Such accommodations to the "machine" created resentment about computers at the workplace. Too often workers were made to feel that the computer was a prima donna that they had to serve. And instead of the computer easing their work, it meant that they had to learn new procedures that created more work for them. The new system often meant more forms to fill out, more filing, more typing, and so on. Just think about the manual work involved in switching over to a seven-digit inventory system from a nine-digit system. After many years with the company, the office workers most likely knew several thousand inventory numbers by heart. Then all of a sudden someone from the outside brings in a "labor-saving" computer and the inventory numbers are changed. Imagine how that made the employees feel every time they tried to look up something by the old number!

Writing new programs or radically changing old manual procedures to accommodate prewritten programs are both bad strategies and should be avoided if possible. Today, you are almost always able to purchase prewritten applications software that you can tailor to your own particular working conditions. The question to ask yourself now is, how can you tell if it's possible to do your project using a prewritten program. What tailorable software exists?

Categories of Applications Software

When you look at tailorable software, you'll find thousands of brand-name programs. If you look closer though, you'll find that most of these programs fall into three major categories—**word processing**, **spreadsheets**, and **data base management** programs. With these three types of applications software packages you can do a wide range of sophisticated business tasks. There are also special-purpose software applications such as billing and accounting programs for dentists and computer-aided design programs for graphic artists. Even these special-purpose application programs are available as off-the-shelf programs, and many of the considerations and much of the vocabulary in our discussion of the three major software categories in Chapters 3–5 is applicable to these special programs as well.

Within each of the three types of applications you'll find hundreds of brand-name programs, but all programs within a category do almost the same thing (despite marketing hype to the contrary). For example, there are many brands of word processing programs available, and all of them essentially do the same thing—**text editing**. I wrote this book using a popular word processing program called WordStar, which was written by programmers at MicroPro. But there are many other brand-name word processing programs available from other software companies such as Perfect Writer, pfs:Write, Multimate, Volkswriter, Spellbinder, Electronic Pencil, Peachtext, and Applewriter, to name just a few. Each of these competing programs has slight differences, but they have all been designed to perform the same type of text manipulation functions that is known as word processing.

There's No One Best Program For Everyone

How do you know which of the competing brand-name products to select? Part of the answer is to have an idea about what features are available so you can figure out which combination of features is appropriate for your particular situation. There is no best program, just like there is no best computer. As always, it depends on what you want the system to do. And this is the most important part of the answer—really knowing what you want it to do. For example, in choosing a word processing program, the needs of a student who will be doing extensive footnoting are much different from someone who's writing a novel, and both of these people need different features from a secretary who is doing mass mailings of standardized form letters.

In Chapters 3, 4, and 5, we will outline the features available in each of the three major types of software so that you can decide for yourself what prepackaged software will best suit your needs. As a general rule, it is easier for you to become computer literate and do your own evaluation than it is for you to hire a "computer expert" to decide what you should buy. In hiring an expert, you must explain the myriad details of your work so that she can understand enough about how you conduct your business to be able to recommend the truly best system for you. Not only do you know the details of your work better than the so-called computer expert, but there simply aren't enough knowledgeable computer experts to go around. For this reason, when you go to a computer store, do not expect the salespeople to know all about the intricacies of the many competing brands of software. Rather, the store personnel will know about the few brands that the store sells (if you're lucky). All too often, people go into computer stores expecting to get educated. Remember, most computer store personnel are salespeople who are hired because they know how to sell, not because they are particularly knowledgeable about computers.

Learning the Vocabulary Is the Key to Self-Sufficiency

Another part of the answer to how to select software is to know the vocabulary. Often software is selected by "experts" at your office and you are expected to learn the programs on your own from the instruction manuals. Unfortunately, software, even the very best, often comes with documentation that is unreadable. I suspect the reason for that is the programming profession seems to attract people who are more comfortable talking to machines than they are to people. And the people who are good at talking to people—the technical writers (who are the most underpaid segment of the industry)—often don't understand the terminology enough themselves to adequately translate it into everyday English. A basic understanding of computer software vocabulary is often necessary to be able to even read the glossy sales literature that advertises the programs. After reading these next few chapters you will understand not only the promotional material, but also be able to shop for software, and actually figure out for yourself how to use the program that you've purchased or are using at work.

CHAPTER 3

Software Applications: Word Processing

Let's begin with **word processing**, since this application is the most popular. Word processing is a program that makes the computer act like a typewriter. The advantage is that this "typewriter" allows you to make corrections faster and easier since you don't have to retype your entire document every time you add or take out material. Let's see what word processing can do for us.

Common Word Processing Features

In order to learn word processing vocabulary, we'll look at old, familiar typewriter-based operations first. You'll see that many of the computer terms that are associated with word processing really stand for the manual procedures that you already know.

Editing

One category of operations you do on a typewriter is called **editing**. You are editing when you are correcting your typing mistakes. For example, you might have misspelled a word because you left out a letter. To correct it on a typewriter, you can use the half-space lever and sneak in the missing letter. In computerese, this is an **insert**. Sometimes a correction entails whiting out a letter with correction fluid and typing the correct letter over the mistake. In computerese this is called **replacing** or **typeover**. Another common correction is when you've typed an extra character by mistake and you need to remove it, but not replace it with anything. In computerese, this is called **deleting**.

Let's look at some examples. Suppose you typed the word "rad." If you really meant to type "read," you have to *insert* an "e." But if you meant to type "red," you have to *replace* the "a" with an "e." If what you intended to type was "ad," you have to *delete* the "r."

These basic functions—insert, replace, and delete—are included in all word processing programs. Just these three functions alone give you incredible flexibility in writing and editing because, unlike a typewriter, the computer lets you insert, replace, and delete as many characters, words, lines, and pages as you want without having to retype your original pages. Unlike the typewriter, where a close inspection will often show where changes have been made, word processing corrections aren't visible at all.

In word processing, changes are made in RAM where all the characters simply get moved in and out of their mailboxes to accommodate your changes. If you insert a new sentence in the middle of page 2 of a five-page document, all the sentences from there until the end of the document automatically get moved down in memory to make room for the new material. When you decide to print your completed five pages, there's no way of telling if a sentence or even whole pages of text were inserted. In the same way, if some characters get deleted, all the characters in RAM get pushed up to close up the space, and there's no telltale hole to show where you whited out the material.

Cutting and Pasting

A function you often perform when you prepare a document is **cutting and pasting**. You may have typed your material and when you reread it realized that your writing would sound much better if you changed the order of some of the

Block of
text to be
moved

FIGURE 3-1. To indicate that you marked a block of text, many computers highlight the block using reverse video. Once you've marked the block you can move it. This process is called cutting and pasting.

sections. You take your scissors, cut out the sections, and tape them together in a different order. If you're very careful when you photocopy the pasted-up version, the tape won't show, giving the illusion that your corrected material is perfect. Thus saving you from having to retype the pages all over again.

With your word processing program, you can also cut out a section of text and paste it in a new location. The way to do it is to **mark** both the beginning and end of the section to be "cut" by typing in a particular command in both places. On most systems, the "cutout" section of text changes color so that the block of text to be moved is highlighted (also called reverse video), as shown in Figure 3-1. Then you merely go to where you want the marked section to be "pasted" and type in another command. Almost magically, the section will appear right where you want it, disappearing from its former location altogether. In computerese, this is called a **block move**. Just about all word processing programs provide you with a block move function. They only differ in the specific procedure you use to indicate what you want to cut and paste.

Search and Replace

Another invaluable feature of word processing is the **search and replace** (also called **find and replace**) function. It lets you scan a document for occurrences of a particular word or phrase. This procedure uses the compare ability of the CPU to find exact matches of a particular sequence of characters. (Experts call a group of letters and/or numbers a **character string**.)

The most common use of the find and replace function is when you've misspelled something or made a consistent error throughout your manuscript. For example, suppose while writing you don't always capitalize "American" and sometimes it comes out as "american." When you've finished the entire report, you can instruct the word processing program to find every occurrence of "american" and replace it with "American." In other words, you tell the program to search for the character string "american." You then indicate a second character string, which is "American" in our example, to be substituted every time a match is found. The word processing program will search your entire document for the first set of characters; every time it finds an exact match, it will replace it with the second set of characters. This search and replace process, in which *all* occurrences of the character string is changed, is called a **global search and replace**.

Another option is the **conditional search and replace**. This word processing function pauses at each occurrence of the character string and waits for you to indicate whether or not you want the characters to be replaced. This feature is used when you want to replace some but not all of the designated words. For example, let's say you are an interior decorator and have worked up a written plan for a color scheme but then decided to change certain gray accessories to dusty rose. You wouldn't want to do a global search and replace, because there are still many places in your plan in which you want to retain the gray decor. By using the conditional search and replace, you could decide at each place where gray was mentioned whether or not to change it to dusty rose.

Follow the Blinking Light

When you're doing word processing you are working at the computer's screen, and without some type of pointer it would often be difficult to locate your place on the "page." To help you, word processing, as well as other programs, provide a pointer, which is actually a square of blinking light on your screen that marks your position. It is called the **cursor**. (In some systems it may be an underline; also, it may or may not blink.) You move the cursor to point to wherever you want to insert, delete, replace, or do block moves. You can make the cursor move around the screen or **scroll**, that is, go forward or backward through your manuscript, by typing in particular commands.

Control and Function Keys

Although there are many commands involved in getting the computer to carry out the word processing operations, such as commands to move the cursor in various directions and commands for underlining, indenting, and centering, there are only two standard ways to give a particular command. Either you use a special computer key, called the control key, in combination with other typewriter keys or you press one special key called the function key.

THE CONTROL KEY

The **control key** was invented to change the meaning of typewriter letter keys so that they could be used to give word processing commands. In this way the CPU can distinguish when it is receiving commands from when it is receiving regular text.

The control key works like the shift key on a typewriter. You know that on a typewriter, to type a capital "D," you hold down the shift key and type a "d." Similarly, to type a control-D, you hold down the key marked control, abbreviated CTRL, while striking the "d" key.

On a computer, when you hold down the shift key to capitalize a letter, the pattern of electrical impulses sent the CPU is different from the pattern that is sent when you strike the same letter key without holding down the shift key. Each key is thus capable of sending two different patterns of electrical impulses. This is how the computer "knows" the difference between small letters ("d") and capital letters ("D").

Holding down the control key makes the typewriter key you're striking send a third set of electrical impulses to the CPU. The CPU recognizes these particular patterns of electrical bursts as commands to perform certain word processing operations and does not treat them as copy to be included in the text.

Each word processing package has its own particular commands to initiate the various operations. Pressing control-D (abbreviated in instruction manuals as "^D") may mean delete in one package and move the position of the cursor forward in another.

FUNCTION KEYS ARE FASTER

Some computer keyboards have special **function** keys on them. These are keys that can be dedicated to doing one particular function, such as sending the command for a block move. Since you only have to strike one key, instead of a control key combined with one or two additional keys, using the function keys can save time. As illustrated in Figure 3-2, these function keys may be arranged in a row above the number keys and are generally marked F1, F2, F3, and so on. Because these markings don't tell you the command for which the keys are being used, it's handy to make labels for them using adhesive tape. Some software companies market self-stick labels or plastic **templates** for you to place above your function keys.

Soft Keys Versus Hard Keys

Some word processing programs allow you to assign your most frequently used word processing commands to the function keys. The IBM PC, which comes with 10 function keys (F1–F10), allows you to do this. If you do a lot of footnoting, for example, you may want the F1 key to mean "footnote here." If you do lots of underlining, however, you may want your F1 key to mean "underline here." When you have the ability to decide which functions are to be repre-

FIGURE 3-2. Some computer keyboards have special function keys. In word processing they are used to perform special tasks such as delete character or underline. Other keyboards don't have function keys. These require you to use the control key in conjunction with a typewriter key. For example, in Wordstar, control-G means delete a character.

sented by function keys, you have what are called **soft keys**, in computer jargon.

This ability to assign functions is in contrast to other systems in which the function keys are permanently pre-assigned to particular tasks by the word processing program. These systems are said to have **hard function keys**.

Some computer keyboards don't have functions keys but may have a way to assign functions to the numeric keypad. In essence, you teach the keypad to generate the codes for particular word processing commands. For example, the Kaypro computer doesn't have function keys, but it comes with a special **configure** program that lets you teach the keys on the numeric keypad to generate specific code sequences. In WordStar the command ∧PS (press control key, while typing "PS"—three separate keystrokes) means underline. On the Kaypro, you can configure the "1" key to generate a ∧PS command, which means that you will need only one keystroke to get an underline. However, if you do this, you must remember that you cannot use your numeric keypad to type numbers. You have to use the number keys at the top of your keyboard, which is just as you would do it on a typewriter.

FUNCTION KEYS VERSUS THE CONTROL KEY

Which works better—function keys or the control key? There is no clear-cut answer. Some people like one system; some prefer the other. What is clear is this: If you only use word processing occasionally (less than once a week), you will tend to forget which combination of keys represents which function in a control key system. Therefore, for infrequent use, function keys are clearly better.

For more frequent use, the choice isn't so obvious. Function keys are usually located on the fifth row of a keyboard, above the number keys (see Figure 3-2) or at the left side of the keyboard (on the IBM PC). This means you have to stretch your fingers quite far from the home keys, something that many people can't do comfortably, particularly without looking. Taking your eyes off your material to locate the correct function keys wastes time. It is relatively easy to get into the habit of holding down the control key, and because it's easy to reach, you don't have to stop and look. However, the control key system requires more key strokes. As you can see, both systems have their good and bad points.

I must confess that my own preference is for the control key. When I learned word processing, computer keyboards didn't have function keys, so I learned to use the control key method. Now I use it instinctively, like a touch typist. In fact, when I try to use a typewriter, I find myself trying to backspace by reaching for the ∧S. Old habits are hard to break. When I use a system with function keys, I find myself going much slower because I'm simply not accustomed to this system—I've lost my "touch."

The Touchscreen and the Mouse

Some computer manufacturers have developed peripheral devices that allow you to move your cursor without having to give typed commands from the keyboard. They are the touchscreen and the mouse.

The **touchscreen** is a CRT that has a grid of tiny beams of light over the screen's glass. When you touch the screen, you break certain light beams. The effect of interrupting the light beam is similar to what happens when you break the light beam across an elevator door opening. In the elevator system, the broken light beam causes the door to open. In the touchscreen system, the broken light beam indicates the coordinates where you want to move your cursor.

The **mouse**, which is a box about the size of a cigarette pack, is used to signal its relative location to the computer. Like other peripherals, it's connected to the computer by a wire. You place the mouse on your desk next to your computer. When you move the mouse toward you, the cursor moves down toward the bottom edge of the screen. When you move the mouse away from you, the cursor moves up toward the top of the screen. When you move the mouse to the right or left, the cursor moves to the right or left, respectively.

Both the touchscreen (popularized by Hewlitt-Packard) and the mouse (popularized by Apple) were developed to appeal to people who can't type. If you feel at all uncomfortable about learning about computers, imagine the feelings of middle-management male executives who have never even learned to type. You may not be able to type well, but most women have an idea of the layout of the keyboard—where the "d" key is for example. Most mothers (like mine) still advise their daughters to take typing "just in case." And, they were right. Being comfortable with an electric typewriter keyboard is one of the major skill requirements in being able to use a personal computer.

After experimenting with the touchscreen and the mouse as alternatives to moving the cursor, I've found them to be overrated. They may be a clever marketing device to make male executives buy these brands of computers, but they are really ridiculous for word processing. Word processing *is* typing, and there's simply no way to get around having to use the keyboard. Using a mouse or a touchscreen to move the cursor means you have to take your hands off the keyboard. This wastes time and doesn't accomplish much. These gadgets, however, are very useful with other applications, especially graphics.

Small Differences May Be a Big Deal

The editing functions we've discussed are available in all word processing programs, but there may be slight variations. For example, one program may have a delete from the cursor to the end of the line command, whereas another program may only have a delete the entire line command. Even these slight differences may be important to you. For example, one poet I know complains that her word processing program will not let her do a block move unless the section to be moved is at least one line long. She likes to move single words around for visual impact and is frustrated that she can't do that on her system. She has to delete the words and reenter them.

From this example you can see that the only way to evaluate a word processing system is to try it out with the type of text you'll be using. Do what you would do if you were buying a typewriter—sit down and type on it to see how it feels. Any computer store worth buying from will let you try out the software on their computer for a few hours so that you can get the "feel" of it. It's just like buying a car—nothing replaces the value of a test drive.

Teaching Yourself Word Processing

In teaching yourself word processing, keep two things in mind. First, there is no need to learn *all* the commands before using word processing. I found that I used word processing very effectively with only the commands for insert, delete, block moves, and cursor movement. Using just those few commands enabled me to get all my writing done. Only after I became accomplished at these did I slowly begin to learn the other features of my word processing software.

Second, most of the commands are not particularly difficult to learn, especially if you are familiar with what the command is going to accomplish. Unless you use your word processing software a lot, you probably won't have all the commands memorized. That doesn't matter, because most of the programs come with a handy reference card you can use to look up the commands that you rarely use. Often the system itself will prompt you along. For example, in Word-Star, when you want to do a search and replace you just look up the command and type it in. Then the program prompts you with "FIND?" When you respond to that question with what you're searching for, it prompts you with "REPLACE WITH?"

The best way to learn how to use a word processing program is simply to try things and see what happens. Taking a class in word processing really isn't necessary. It's very much like learning how to type. The way you learned to type was by practice (you typed). Listening to someone tell you about typing doesn't help much. Neither will a 4-hour class in word processing. The reason you took typing in school was to practice and to have a teacher grade your work. But the computer is an excellent teacher. If you do something wrong, *it* will tell you!

To avoid frustration and headaches when learning word processing, use the word processing program for no more than 1½ hours at one sitting. Breaking up your learning time into four or five separate sessions (on consecutive days) is a more natural way to learn the commands than to try to master the program in one long day. Naturally, there will be some frustration involved in learning, but by approaching it in small doses, it will be minimized. And after you've learned it all, you'll be surprised at how easy it really is.

Optional Word Processing Features

So far we've discussed features that are included in almost all word processing programs. Now, let's look at optional functions that may or may not be included in a particular word processing package. Sometimes these functions are available only as options that you will have to pay extra for. Sometimes they are simply not available for a particular program, especially for word processing programs for home computers.

Optional functions include footnoting, page break indicators, indexing, spelling checkers, thesaurus program and merging. Footnoting is a good example of why software companies include optional features. Why should you pay for this capability if you are never going to use it? By separating out the less popular features, some companies have been able to keep the prices of their standard packages down.

Page Break Indicators

One reason I particularly like WordStar is that on the screen it shows you exactly where the page breaks will occur when the document is printed out. This helps avoid pages that have only one line or one word (typists call this a "widow"). Other word processing programs, such as AppleWriter, don't indicate page breaks so you don't know that a page looks bad until you've actually sent your text to the printer. If it isn't right at that point, you have to throw away the pages, reedit your text, and print out the pages all over again. This is quite time consuming and wastes expensive stationery.

Indexing

The indexing option allows you to specify key words, which the software then uses to generate an index, citing each page number in the text where the word occurs.

Computer indexing is based on the computer's ability to compare two things to see if they're equal. Building the index just means that the program takes one word at a time from a list you give it and compares it to each word in your text. Every time the word from your list is the same as a word in the text, the page number is printed. Of course, if you've misspelled a word in the text, it will not appear in the index.

Indexing is a good example of a job that people used to do that is being drastically changed by computers. Of course, if you've ever used a book with a computer-generated index you may have noticed that the citations are sometimes quite stupid, including references to words that may have only appeared in passing on a particular page. (So citing it as a reference was actually misleading.) Thus, for good indexing, people must mark each word to be cited to indicate that the particular references are meaningful. Computers have changed the amount of time needed to index a book properly and changed the way indexers do their work—but they have not eliminated the need for indexers.

Dictionary Programs

Spelling checkers, also known as dictionary programs, use the compare function, too. These programs are made up of two parts—a program and data. The data are common words (without their definitions). The program checks every word in your text against its list of words. If there's a word in what you have written that doesn't appear in the dictionary data file, that word is automatically highlighted in your text. This gives you a chance to verify if the word is indeed misspelled or if its simply not common enough to have been included in the dictionary data file. Good spelling checkers come with a data file of between 30,000 and 75,000 words.

A good spelling checker program will also allow *you* to add your own words to the dictionary program. Typically, you can add 3,000 to 5,000 new words. When you run the spelling checker against your text and a word is highlighted as a suspected misspelling, you can either correct it, add it to the dictionary, or leave it as is. As you add new words to the data file your own vocabulary begins to become part of the dictionary. This is especially handy if you are in a specialized profession such as law, medicine, or even computers. After a while, as you've added the special words (or even proper names) you most commonly use, you'll find that when the spelling checker highlights words in your text, the chances are good that they really are misspelled.

When a word is highlighted as a suspected misspelling, you can request to have a list of possible correct spellings displayed on your screen. You can then choose the one that's right, and it will automatically get inserted into the text without you having to type it in yourself.

Spelling checkers can only check spelling. They do not check grammar or meaning. There is no way they can tell if you've used the word "two" instead of "too" or "principle" instead of "principal." If you typed in "I red a good book," the spelling checker program could not identify this as an error.

Thesaurus Programs

Thesaurus programs are similar in operation to spelling checkers. With a thesaurus program you move the cursor to a word and then request a list of alternate words. This is really not much different from how you would use a printed thesaurus, except it's handier.

Merging

An important option you can get with word processing is the ability to do **merging**. Merging is the way that computerized form letters are created. There are two types of merging—paragraph assembly and form letter merging.

PARAGRAPH ASSEMBLY

Paragraph assembly is what is usually referred to as "boilerplate." You type standard sections of text, perhaps paragraphs, and store them on disk, identifying each by unique file names. When you are creating a document, you call up the desired paragraph by merely typing its name. As many paragraphs as you want can be called up and merged together to form the finished product.

The letters you receive from your congressperson are created by paragraph assembly. Your personalized letter may contain the standard opening paragraph thanking you for writing, the MX missile paragraph, the aid to education paragraph, and the current month's standard closing paragraph, which may remind you to vote in the upcoming election. Once the standard paragraphs have been entered, the only typing required to generate such a letter is the characters representing the names of the paragraphs to be merged together. For example, the congressperson's assistant would have typed SO, MX, ED-AID, and SC. Text merging also allows you to preset pauses, or blank spaces, in which you can type special comments, people's names, and so on. Thus, the SO paragraph probably has a preset pause to allow the typist to enter your name.

Other professionals also prepare documents using the paragraph assembly approach to merging. For example, law offices prepare wills and contracts, government consultants prepare proposals, and contractors prepare bids this way.

By using the paragraph assembly option to word processing along with the preset pause capability, you can create standardized forms such as leases and insurance claims forms easily. All the boilerplate that's customarily required, such as the physician's name and address on insurance forms and any standard language, is preset. In addition, the tabbing and carriage returns that you normally have to do on a typewriter to get preprinted forms to line up correctly are preset, eliminating much of the time-consuming hassle of manually typing information onto forms. The cursor automatically jumps to the next blank space, and the typist merely has to enter the appropriate "personalized" words or phrases, after which the cursor moves on to the next blank.

FORM LETTER MERGING

Form letter merging is used to add individual names and addresses to a standard form letter. For example, you may want to send the same form letter to every client on your mailing list. By using form letter merging, every letter will be printed out as an original; it's impossible to tell that they weren't all individually hand typed. Unlike preprinted or photocopied form letters, there are no telltale empty spaces or differences in the typeface where the recipient's name and address has been typed in.

Form letter merging is easy. All you have to do is type in your list of names and addresses and the merging program will automatically print out letter after letter, with each one having the next name and address on your list. When merging mailing lists, you can also choose to merge text, such as the street name or the person's name, into another place in your letter. In this way the letter will seem personally written. For example, merging in the street name can generate a letter that includes a statement such as "all your neighbors on Maple Street . . . ." Other information that you maintain on your computerized mailing list, for example, an expiration date, a pet's name, or the date for an annual checkup, can also be merged into the text of a form letter (which is how you get those nice reminders about coming in for your annual dental examination).

Less expensive word processing programs generally don't provide text merging. After all, if you're doing word processing on a home computer, you're probably writing letters to friends or editing school reports, not generating 5,000 personalized form letters.

Is What You See What You Get?

In addition to the optional features we've just discussed, a big difference among word processing programs is whether or not what you see displayed on your screen looks exactly like what you see on a **printout**, that is, the printed material produced by the computer. Generally, a CRT screen won't be able to display an entire page. Most business computers are set up so that the screen shows you a maximum of 24 lines with 80 characters on a line. For most people (including me) this is perfectly adequate.

Some people, however, want the screen to display the entire 50 or 60 lines that appear on a printed page. To do that requires a special monitor (screen) and a word processing package that is programmed to be able to use the extra screen length. You could also add a special controller board that generates smaller characters on the screen for the same effect. Other people may have different display requirements, such as an extra wide screen. They may use the DEC Rainbow computer, which comes with a screen that will display up to 132 characters per line. Again, such special requirements generally entail a special peripheral device, as well as word processing software that is tailorable for this kind of expansion.

At the other extreme are the home computers that use a 40-character display screen. As we pointed out in Chapter 1, with that type of display there is no correspondence between what you see on the screen and what your printed copy looks like. The original Apple II is one well-known brand of computers that uses a 40-character display. A later model, the Apple IIe has been expanded to include an 80-character display (the "e" means "expanded").

Installation Menus

If you decide that you need a special extra long or extra wide monitor, most word processing programs will have to be tailored to work with it. Tailoring, or **configuring**, the package is a simple operation if the monitor you want to use is included in the *list* of monitors in the word processor's installation program.

Let's diverge here for a moment. The type of list we just mentioned actually appears quite often in packaged software. As with all computer devices, it has a special name, which is a **menu**. A menu is a listing of choices that the computer gives you. All you do is type the letter or number next to the appropriate choice, and the computer takes it from there. Think of it as a multiple choice question, except that you can usually change your mind.

When you install your word processing program, one of the menus you see is a CRT menu. If the monitor you wish to use is one of the choices, you merely type in its selection number. If the monitor you want to use is not listed in the menu, it may be quite difficult, or even impossible, to make it work with your word processing program. Even tailoring has its limitations.

Dedicated Word Processors

The term **dedicated word processor** means a computer that is dedicated to doing only one task—word processing. Although it is clearly a computer, it is restricted to only being able to run one program—the word processing program that it came with. It can't run spreadsheets, accounting programs, or other common office applications. At one time, dedicated word processors had much more sophisticated word processing programs than were available for personal computers, which accounted for their appeal. But those days are passing. For example, Wang word processors have been considered by many users to be the "Cadillac" of dedicated word processors. But it's now possible to purchase a word processing program called Multimate that makes the IBM PC act identical to a Wang word processor.

Why buy a dedicated word processing machine if you can buy a personal computer? That's a good question, especially since dedicated word processors cost more than personal computers—often two or three times as much! That just shows you what the miracle of marketing and advertising can do—it can get people to pay more for a machine that can do much less. Another reason is that many of the middle management executives who authorize the purchase of this type of equipment know less about computer technology than you now do.

There is also a subtle type of exploitation going on. Word processing marketing executives play on the fears that many secretaries have about computers. Instead of trying to sell a computer to the executive secretary, the company will sell an electronic typewriter or a word processor. The promotional material designed to appeal to the secretary never mentions that what she's really buying is a computer. Instead of trying to educate secretaries, computer companies exploit their fears of the unknown (the computer) and sell them a less powerful machine at an inflated price.

Printing Out Your Text

Many people purchase dedicated word processors instead of computers and word processing programs because they believe (mistakenly) that this is the only way to get high-quality printed output. The truth is that the quality of the printed page, known also as **hard copy**, is dependent only on the type of printer you plug into the computer.

There are two major categories of printers—dot matrix and letter quality—available in the consumer market. Other more expensive output devices are available for industrial use or special purposes such as typesetting or microfiche. Laser printers, which use technology similar to photocopy machines, are just beginning to drop in price and enter the consumer and small business market.

Dot Matrix Printers

The **dot matrix printer** uses dots to generate printed characters in the same way that some scoreboards use lightbulbs to show the name of the home team or the score. The more dots used per square inch to make up the characters, the better the quality of the printed text (Figure 3-3 on next page).

Dot matrix printers were invented to provide a high-speed, yet inexpensive method of printing. They print faster than electric typewriters because there are fewer moving parts. Unlike electric typewriters that use a ball element that pivots in two directions, dot matrix printers use a stationery printhead. Having fewer moving parts also makes dot matrix printers less expensive.

The more sophisticated dot matrix printers are designed to run at two or more speeds. The slower speed can be used for finished work; it produces so many dots per square inch that

FIGURE 3-3. Dot matrix printers use dots to generate printed characters. This is similar to the way that some scoreboards use lightbulbs to show the name of the hometeam. The more dots used per square inch to print characters, the better the quality of the printed text.

it's difficult to tell that the material wasn't printed on a regular typewriter. The faster speed produces fewer dots per square inch, and the output looks very "computerish." This kind of output is fine for first drafts, mailing labels, and in-house records.

Letter-Quality Printers—Slower but Sharper

Letter-quality printers are exactly that—they produce the high-quality print you're used to seeing on business letters. The text looks exactly like it comes from a good electric typewriter using a carbon ribbon. In fact, it prints in a similar way as an electric typewriter, only instead of using a ball element it uses a wheel with the characters imprinted on it.

The wheel is an improvement over ball elements, because

it rotates faster and only in one direction. It still, however, takes time to spin to the appropriate letter, which makes it slower than dot matrix printers. Since the print wheels used in letter-quality printers look like a daisy, they are called **daisy wheels** and the printers are generically called **daisy wheel printers** or **daisy printers**. Like the old-fashioned balls, the daisy wheels come in variety of typefaces and sizes.

Printer Speeds—How Fast Is Slow?

Printer speeds are usually given in characters per second— abbreviated **CPS**. Since we're used to thinking about typing speed in words per minute let's start with that and convert it to characters per second.

If I type 60 words per minute, it's the same as typing one word per second. An average word is about six characters long, so I type six characters per second, or 6 CPS. An excellent manual typist may type 7–9 CPS.

Dot matrix printers range from about 100 to 200 CPS or 16 to 33 times faster than I type. Let's be conservative and look at the 100-CPS dot matrix printer. It types about 960 words per minute!

Now let's consider letter-quality printers. The fastest letter-quality printers run at about 50 to 60 CPS. Since I type 6 CPS, the fastest letter-quality printers go 10 times faster than I type. For less money, you can get letter-quality printers that operate at 15 CPS, or twice my typing speed of 60 words per minute.

Which Printer Do You Really Need?

In evaluating printers, one of your main criteria will be the amount of material you expect to print out each day. The more material you will be printing, the faster your printer will need to be, or it won't be able to keep up with the workload. In order to compare printers, let's begin by figuring out how long it will take particular printers to print out a page.

If your margins are set at 15 and 85, you have 70 characters on a line. A standard letterhead-sized sheet of paper has about 55 lines on it, giving 3,850 characters. Let's round this down to 3,000 characters to allow for top and bottom margins and blank lines (and to make our arithmetic easier). With the fastest letter-quality printer going at 60 CPS, it will take 50 seconds (3,000/60), almost a minute, to print a page. This high-speed daisy wheel printer costs about $3,000. A less expensive daisy wheel printer, one that costs about $600, will type about 15 CPS but will take 200 seconds (3,000/15), or more than 3 minutes to print a page. The slower letter-quality printer takes 5 hours to print a 100-page document. Even the fastest letter-quality printer takes almost 1½ hours!

Contrast this with dot matrix printers. A 100-CPS dot matrix printer costs only about $300 and will print our page in about 30 seconds; thus a 100-page document takes about 50 minutes to print. A dot matrix printer that prints 200 CPS will print it in half that time and costs about $800. (I hesitate to put prices into print because they keep going down. I did it here to illustrate relative values of printers and to show that dot matrix printers give you a lot of speed for your money.)

Recently, the quality of dot matrix printing has improved enormously. You have to look very closely to tell that the text was printed using a computer. There is, however, incredible resistance to even this good-quality dot matrix printing. Just as electric typewriters with carbon ribbons are the current business standard, the daisy wheel printer is the standard for business printing. People still want to believe that they are getting personal letters from their congressperson. It's ridiculous, but there it is. Maybe people are afraid to trust a company that can't afford a daisy wheel printer.

Whatever the reason, offices are still insisting upon these printers.

Many schools also require letter-quality printers. These schools insist that masters and doctoral theses be printed on letter-quality printers. This situation is reminiscent of when typewriters first came in and there was great resistance to receiving typewritten correspondence because the business standard was handwritten letters!

The protest against dot matrix printers will probably change in the foreseeable future. Printer prices are about as low as they can go, so the competition among brands has changed to how "pretty" the dot matrix output is. With the very newest printers you need a magnifying glass to see the dots. Not only has the print quality been improved, but it is now possible to use different fonts (typefaces) without having to change the printhead element. The instructions for how to arrange the dots into a given font style are now stored in memory so that a simple command brings up another typeface (Figure 3-4).

Newer printer technology using this approach includes ink-jet printers, which "shoot" dots of ink, and laser printers. Since these printers don't have a print head that strikes against a ribbon, they are referred to as **non-impact printers**. Because there's no impact, they are extremely quiet. Because laser printers have no moving printhead, they are

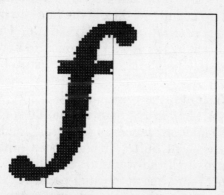

FIGURE 3-4. In correspondence quality, ink jet, and laser printers, fonts are generated by particular patterns of dots. These font definitions are stored in memory, permiting different fonts to be used in a document without having to change the printhead. (Courtesy of Compugraphic Corporation.)

extremely fast—with speeds measured in pages per minute.

Print Buffers Regulate the Flow of Characters

Another reason why printer speed is important is that during the time that printing is being done, the computer is occupied and unavailable for other work. Imagine this. It takes 5 hours to print out a document and a client calls up with a question just as you begin printing. You would be unable to look up her computerized records until the printing was finished; that's almost 5 hours.

No matter which kind of printer you use, its speed is snail-paced in comparison to the speed at which the computer's CPU is able to send characters out to be printed. In order for these two pieces of equipment to work together, there has to be something to regulate the speed at which the printer receives characters from the CPU so that the poor printer doesn't get hopelessly overwhelmed. **Print buffers** were invented for just this purpose.

To understand how print buffers work, visualize a bathtub being filled up at full speed from the faucet while at the same time water is trickling out of its drain. Depending upon how slowly the water is trickling out and how fast the water is pouring in, the water level in the bathtub will rise or fall.

As Figure 3-5 shows, the print buffer is a special area of RAM that fills up, similar to a bathtub, with text characters from the CPU. The word processing program sends characters of text to the print buffer at high speed (imagine the water faucet turned all the way on). Because the buffer is silicon-based RAM and has no moving parts, it can receive the characters as fast as the CPU can send them. Simultaneously, a special controller sends the characters in the print buffer to the printer at a slow enough speed for the printer to handle them. The speed at which the characters "trickle out" of the buffer depends upon the type of printer you

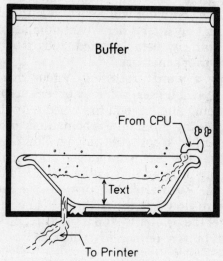

Buffer

From CPU

Text

To Printer

FIGURE 3-5. A print buffer is an intermediary between the high-speed computer and the much slower printer. To understand a buffer, think of a bathtub being filled with characters (water) at high speed, while characters (water) are draining off slowly to the printer. The buffer (bathtub) never overflows. If the level of characters (water) reaches the top of the buffer (bathtub), the tap is turned off and characters stop being sent to the buffer from the CPU until enough of them are drained off by printing.

have. A 50-CPS printer will accept characters twice as fast as a 25-CPS printer.

In our analogy, the height of the water in the bathtub is the number of characters being held in the print buffer. The amount of text in the buffer (much like those old math problems from high school) can be determined by the difference between the rate of characters going to the buffer and the rate of characters trickling out to the printer.

There is one difference between the bathtub example and the print buffer. If the print buffer fills up (because the document is much larger than the buffer), it will signal the CPU to stop sending it any more characters for a while. Unlike a bathtub, the print buffer will never overflow—no text will ever be lost. The computer will resume sending characters to the print buffer only after a significant number of characters has "drained off" by being printed, thus freeing up more space in the buffer's memory.

When using a word processing system that includes a printer with a print buffer, you may experience the eerie situation of noticing a mistake in the typed output and giving the word processing program the command to stop, only to find the printer just keeps typing away. This event is caused by the fact that the text is no longer under control of the word processing program; it's in the printer's buffer. Think of our bathtub. Even after you turn off the faucet, water will continue to flow out the drain until the tub is empty. The way to get the printer to stop is to turn it off or to hit the buffer's **reset button** (either of which causes the buffer's RAM to "lose" its contents).

Most printers come with a small print buffer. You can buy larger print buffers as a separate component to upgrade your system. If you have a very large print buffer—one that can store many pages of text—your computer will be freed up to do other jobs once all the characters to be printed have been sent to the print buffer by the CPU. If you have a 512K print buffer that holds about 170 pages (512K/3K per page), you can use your computer almost immediately after giving the print command, even though your letter-quality printer will be printing all day.

SPOOLING

Another approach to printing is called **spooling**. Spooling is a *software solution* to the problem of freeing up the computer for use. With spooling, the material to be printed is sent to the computer's disk drive instead of directly to the printer. A special spool program directs the CPU to read the disk and send characters to the printer. The spool software takes advantage of the fact that the CPU is often idle (while waiting for you to type, for example). Although spooling frees up the computer for new work, printing is slowed down considerably since characters are only sent to the printer when the CPU isn't occupied elsewhere. The hardware approach of a printer buffer is therefore preferred to the software approach of spooling when large quantities of printing are required.

Printers and Your Word Processing Software

As with extra wide or long screens, you must tailor your word processing program to work with your particular printer. Simply call up the **printer menu** of your word processing program. To install a printer that is listed in the menu, just press the number next to it. To install a printer not listed in the menu may be difficult or even impossible.

As you might imagine, there's a lot of competition among the newer peripheral manufacturers to get their products included in the menus of the popular word processing software. To ensure that their printers will be compatible with the software, a common approach is to design new products to **emulate** or mimic printers that are already listed in the menus.

For example, one of the earliest companies to make printers was Diablo; therefore, the Diablo model printers are included in almost all word processing printer menus. Some of the newer companies, such as the Japanese manufacturer Brother, sell printers that emulate Diablo. In their sales literature they may say, "Choose the Diablo 630 model printer on printer menus when installing this Brother printer."

Be cautious; not all emulations work correctly. Make sure that your computer store has had experience installing your particular printer or that they will take back the printer if you cannot get all its features to work with your word processing package.

Typewriters Are Not Printers

Printers are designed and created for only one purpose—to work in conjunction with computers. You have probably seen advertisements for electronic typewriters that can be hooked up to computers. Unlike printers, they have keyboards and are designed primarily for manual use as typewriters. They may have attracted your attention since they cost as little as $350; but do not be fooled. They are not a good idea for day-in and day-out use as a computer peripheral. When they are used as computer printers, they are driven at maximum speed (about 10 to 12 CPS) continuously, which causes them to break down frequently. In fact, one way printers are evaluated is by their "mean (average) time between failures" (abbreviated **MTBF**) and these pseudoprinters have notoriously bad MTBF ratings.

Another thing to keep in mind is that the price of these electronic typewriters is deceptive, since it does not include the controller board. You can't just hook up an electronic typewriter to a computer and expect it to work. You need a controller to convert the computer's binary codes into commands to strike the correct key on the typewriter. The controller board, when sold separately, is called the **interface** and may cost $100 or more.

There is an interface that allows you to hook up an IBM Selectric typewriter to your computer for use as a printer. Don't purchase it unless you only plan to print a few pages occasionally. Not only will the Selectric wear out very quickly with even moderate use, but because of its ball element design, its top speed is less than 10 CPS.

New Technologies from Word Processing

Word processing is the foundation of many exciting computer-based applications, such as electronic mail and typesetting, that are causing a revolution in printed communications. The fact that technology will have a great impact on the "office of the future" is obvious, and although there is much debate over exactly what this will mean, everybody agrees that we are in a time of profound change, and that white-collar work will never be the same.

Electronic Mail

Letters and other documents that you've written using word processing can be transmitted via a modem to other computers. As you may recall from Chapter 1, a modem is a device that enables you to use your telephone to send messages via your computer. This is known as **electronic mail**. The phrase "electronic mail" means different things to different people. In general, there are three approaches to electronic mail.

COMMUNICATIONS PROGRAM

First, suppose you want to send an "electronic letter" to someone who has a computer. No mail carrier or courier is used. Instead your "letter" is transmitted from one computer to another using your modem and a **communications program**. Basically, the communications program reads your letter from the disk on which it is stored and sends it to your modem, which is connected to your telephone line. As soon as you dial the phone number of a friend or co-worker, your letter is electronically transmitted over the telephone lines to her computer. When the incoming letter arrives at her computer (via her modem and communications program), she can direct it to the screen, the printer, or the disk. She'll probably receive it on her disk to speed up the process (the printer is much slower than the disk) and save on long distance telephone charges. Later, she can read it on her screen or, if she wants hard copy, print it out.

The drawbacks with this method are that the recipient's modem and computer must be turned on, and a communications program must be loaded into her computer's RAM in order to pick up the call when it comes in.

ELECTRONIC MAILBOXES

What do you do if your friend isn't home, or the phone is busy, or her computer is tied up doing something else? An alternate approach is to use an intermediary such as The Source or CompuServe. These are information utility companies that have large central computers. Subscribers to these services rent an **electronic mailbox** which is actually tracks of space on the enormous disk drives used by the utility company.

To send your letter, you call a utility company such as The Source (your friend also has to have an account with them) and transmit the letter to your friend's account. You need to know her account code, which is similar to knowing a company's telex identification code. When your friend gets home, she calls up The Source and reviews the letters that have been recorded there.

An added advantage to this mode of electronic mail is that you don't even need to own a computer. Since the transmission program is in the utility company's computer and the letter is stored on their disks, all you (and your friend) need is a terminal and modem—equipment that now costs under $200.

As you might imagine, electronic mail is making the technology of telex machines obsolete. Western Union, the company that sells the telex service, has realized this and has begun offering its own electronic mail service under the brand name EasyLink. EasyLink is just becoming international. For example, Tina International Message Service allows personal computer users to send an instant electronic letter from New York to Paris for about 5 percent of the cost of a conventional telex.

A similar system of electronic mailboxes is used by many major corporations. Instead of using the disk at utility companies such as The Source, they use their own central computer's disk storage. Employees using personal computers (or terminals) send letters or memos to each other via electronic mailboxes on their company's disk. This saves all the time normally wasted on "telephone tag" (you call and leave a message, then the other person calls you back and leaves a message, and so on). With electronic communications, you transmit your exact information to the person, in your own words, without worrying if it got garbled by the message taker. Sophisticated electronic mail systems even let you verify the date and time the person reviewed the letter and you can rate the priority of the letter from urgent to routine. Best of all, they let you send multiple copies, perhaps to all the branch managers telling them of the next scheduled meeting.

ELECTRONIC MAIL WITHOUT COMPUTERS

The third type of electronic mail is designed to let you send letters to people who don't have computers. The main appeal of this approach is speed—recipients get their letter in one day or less.

Mail companies have remote printers located around the country. You call a nearby service center and transmit your electronic letter using your communications program and modem. The letter is forwarded electronically (via the mail company's modems and the telephone lines) to the printer nearest its final destination. Then it's printed, stuffed into a window envelope, and delivered, using either the U.S. Postal Service or a private courier service.

One company offering this service is MCI. It has 16 remote printing sites and offers a variety of rate structures. For example, MCI has contracted with Purolater Courier for messenger service that guarantees four-hour delivery to selected cities. A less expensive rate provides overnight delivery. A still less expensive rate can be had by using the U.S. Postal Service to deliver the letter locally. Western Union has announced a similar plan to deliver EasyLink mail internationally using DHL Worldwide Courier Express.

Computerized Typesetting

Typesetting is another industry that is being affected by word processing. Until now, you typed the material you wanted printed, proofread it, and took it to a typesetter. The typesetter retyped your material, usually introducing typographical errors. You then had to proofread your material again and correct any mistakes. The typesetter made the indicated corrections, and the corrected copy was photographed into the form required for printing—camera-ready copy.

But modern typesetting machines are actually computers that have a special output device—a camera. Therefore, instead of paying typesetters to type your material, you can now electronically transfer your text from your disk to the typesetting machine. You can do this by giving the typesetter your disk or transmitting your material to the typesetting machine via a modem. Since there are usually problems with compatibility between different types of disk formats, the modem approach is more common.

This new technology may have revolutionary effects on the whole publishing industry. By transferring a manuscript from disk to camera-ready copy without having to retype it, the cost of publishing will drop dramatically. This should make it much easier for people to publish their own work. As book publishing becomes more commercially oriented, this may be the only way we get to see books that aren't the mass appeal supermarket-rack variety.

National Newspapers—A Result of New Technology

You can already see the impact of the new technology at most daily newspapers. Reporters now enter their stories using word processing. The newspaper is then "pasted up" electronically, using block move commands and other special formatting operations, and printed without having to be typeset. With this technology, newspapers no longer have to be limited to one locality. Several national newspapers in fact have taken advantage of computerization. Both the *New York Times* and *USA Today* use modems to transmit their "pasted-up" newspaper pages to remote printing sites. The Los Angeles and New York City editions of the *New York Times* are printed simultaneously in their own locales. This eliminates the delay and extra expense that resulted from printing in one location and shipping newspapers across the country.

Office Automation— New Type of Factory Work

Word processing is an exciting tool. It changed my life! It has enabled me to write this book. I used to agonize over every word I wrote, hesitating, scratching out, throwing away pages, and starting over again. But now, using word processing, I just write, without worrying about it coming out perfectly. I know how easy it is to go back and edit later. My creativity has been liberated.

Word processing may even prove to be a boon for children who have difficulty with writing by hand. Sometimes the problem is motor control, sometimes it's other disabilities such as dyslexia. In any case, experiments are now being done in teaching those children word processing, and the results seem quite promising.

Unfortunately, everything about word processing is not good, particularly its use in the workplace. There are offices with **keystroke monitoring**, where the right to stretch and take a rest from eyestrain and back fatigue is denied some workers. This problem is part of the larger problem of using office automation to convert office work into factory work.

Automation is affecting all service industries. Think of fast food "restaurants." They're just food factories, really. In the secretarial area, office automation often turns a relatively interesting job into a repetitious, piecework job. In many offices the introduction of word processing means that jobs may be segregated into first-draft typing and revision typing. One typist does only first drafts; another typist does only editing. A typist no longer feels that the finished product is hers. Any satisfaction and pride she used to feel in knowing that she was responsible for creating professional-looking documents is gone. Like the autoworker, she is a pieceworker.

The "efficiency expert" who brings in the first draft/revision system doesn't figure in the costs of worker dissatisfaction, which may only be evident by the extra sick days that typists start taking. The keystroke monitoring system proponents haven't factored in the medical costs of eyestrain either, and, in fact, the first worker's compensation claims are just now being awarded to CRT operators.

I hope you see that it isn't the computer or the word processing system that is at fault, it's the method of implementation. I believe that understanding this puts you in a better position to speak up for decent working conditions when a computer-based system is proposed for your workplace.

CHAPTER 4

Software Applications: Database Management Systems

The term **database management system** sounds impressive, highly technical, and intimidating. So what else is new? I call it "electronic filing," which just as aptly describes it. But of course, consumers are more likely to pay more for "database management software" than for a program to do electronic filing, so you don't usually see the term "electronic filing" referred to in computer publications.

Once you understand database management systems, the information utilities with their electronic filing cabinets stocked full of information will become accessible through a mere telephone call. You'll be able to "look things up" using these services. Understanding electronic filing systems also means that you'll have the ability to set up your own electronic filing cabinets filled with information that you've collected for yourself. You will be able to get rid of the zillions of scraps of paper that we all seem to collect by storing all that information in an orderly and useful way—in your electronic filing cabinet. This can even turn into your own business if your electronic information file, called a **database**, is useful for other people. As soon as you learn the fundamental concepts of database, we'll brainstorm about other applications and I'll explain my own database information service.

To understand database systems, here's a review of the manual systems you're already used to. By applying computer terminology to those familiar operations, you'll see how much you already know about database management.

What Is a Database Management System?

Visualize a manual filing system that uses 3 × 5 cards to keep track of the members of a club. Every time someone joins, you fill out a card with the appropriate information, such as name, address, the phone number, and date of membership. Whenever anyone moves, you have to pull her card and change the address and perhaps telephone number. If you needed to call a member, you'd pull her card and look up her telephone number. Periodically, perhaps every month, you type address labels for the club's newsletter and sort them into zip code order so you can save money on postage by using bulk mail. Perhaps every other month you look through the file to find the people whose memberships are up so that you can send them renewal notices.

Computerized Filing

Now let's look at these same filing system operations when you do them using a database management program. Instead of filling out a card, you type the information into a computer. This process is called **data entry**. When you change the address on a card that is in your file, you do an **update**. When you look up information like a telephone number on the card without changing it, you perform a **query**. When you pick out the people who are up for renewal, you do a **selection**. And when you put the addresses into zip code order, you do a **sort**.

As you might imagine, database management is one of the most useful applications of computer technology. Every time a magazine changes your address on its records, or the police stop you for a traffic violation and run a computer check on you, or you use your credit card or you make a plane reservation, you are experiencing the use of database management programs.

Learning how to use and tailor database management systems will make your computer much more useful and will make a huge amount of information available to you.

What Makes Up a Database?

Before looking at database management systems in detail, we need to become familiar with some database vocabulary. It's easy.

Records

Instead of using the word "form," computer specialists use the word **record**. Thus, each 3 × 5 card in our membership file is called a record in computer terminology.

We often think of a record (form) as holding information about an individual, like the member of a club, or a student's scholastic record grades, and people's credit ratings. But records are not limited to keeping track of people, they can keep track of cars, or real estate, or anything. For example, the Department of Motor Vehicles has a record for every registered car, and the property tax collector has a record for each house (the owner's name is only an incidental piece of information that changes every time the property is sold).

Fields

Since the beauty of electronic filing is that you can use the computer's incredible speed to look up or process information that's contained on records, there needs to be a way to identify which piece of information on a record is the one you want the CPU to deal with. For example, if you want to sort all your membership records into zip code order, the CPU needs to know how to avoid confusing the zip code with the street address or phone number.

To do this, each type of information on a record is separated into its own section. We do the same thing on paper forms by using blanks. For example, our club membership form (record) has a name blank, an address blank, a telephone number blank, and a date of membership blank. In database terminology the word for blank is **field**. As illustrated in Figure 4-1, our electronic 3 × 5 cards have a zip code field, telephone number field, a date of membership field, and so on.

Files

We're all used to the word **file**—it's the manila folder that you keep papers in. Computerized database systems also use

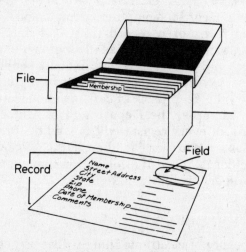

FIGURE 4-1. Computer specialists call a form a record, a blank a field, and a collection of records a file. This 3 × 5 card is a record with 8 fields.

the term file quite similarly, to mean the thing that holds your records. It's important to know that database systems require that every record in an electronic file has the same layout as all the others. In other words, the records in a file must all have the same number of fields assigned to hold the same type of information. For example, one file can hold all the club's membership records, but another file is needed for the club's budget records since they are kept on a different type of form.

LARGE FILES REQUIRE LARGE DISKS

Most programs require that your entire file be on a single disk; therefore, a large file size means you'll need a large disk. Often the size of disk you need will dictate which computer system to buy. For example, an Apple IIe uses disks that have one-third the disk storage capacity of a Kaypro 4. If you have a file of 250,000 characters, you could use the Kaypro 4 but not the Apple IIe. It is possible to upgrade the Apple IIe by plugging in a larger capacity hard disk drive (at a substantial cost). With certain brands of computers, however, there is no way to upgrade. For example, you will have real problems if you buy a Sanyo without figuring out your file size first since the Sanyo's disks hold only 180K and there's no way to upgrade the system.

CALCULATING FILE SIZE

When you're deciding which computer system to buy, the first thing you must calculate is what your *maximum file size* will be. Remember, you should be analyzing your software needs before you ever buy your hardware. To help you determine the size of your file, let's look at some examples.

Suppose you have a regular filing cabinet containing membership information. Your organization has 800 members, and you therefore have 800 records. Now suppose you are a survey researcher and you are conducting a 10-page survey of 800 people. You will also have 800 records but each will be 10 pages long. Although both files have 800 records, when you look in your filing cabinet, you'll see that the survey file takes up much more room than the membership file. When calculating your file size, you have to consider the size of each record, as well as the number of records.

Similarly, when calculating the size of an electronic file you don't only measure it by the number of records that it contains. Rather, you count the maximum number of characters in a record and multiply that by the number of records in the file. For example, the club membership record may have *fields* for name, street address, city, state, zip code, telephone number, and renewal date, and the total *record* length may be 100 characters. The *file* size will then be 800 records × 100 characters, or 80,000 characters. This means that the disk capacity you need to hold your entire file is only 80K. On the other hand, each of the survey forms may contain 1,000 characters. The 800 completed surveys will therefore require a computer with at least 800K (800,000 characters) disk storage and will require more careful hardware shopping.

Most computer stores will have someone to help you figure out the size of your file, but you must be able to provide the basic information such as how many clients you have and how much information you want to keep on each client. Specific details such as the number of characters you want to use for each field are also necessary for the calculations.

Just like everything with computers, there is a special name for this process of figuring out the size of your file; it's called **systems analysis**. The person doing the figuring is called a **systems analyst**.

Teaching the System about Your Data

Before you can use your database system and begin typing in all your information, you must first tell the system what fields you'll be using and how many characters each field can hold. What you are actually doing is tailoring the generic database software to your own needs; this process is known as **installing the system** or **initializing the system**. The system needs this information to figure out the total length of your records. After you describe your data, the system will be able to locate individual records from within a file.

Let's say a particular file holds records containing 300 characters each. When you select record number 835, the computer simply multiplies 835 × 300. This tells it how many characters into the file your particular record is located. The read/write head can then be directed to the correct track and sector on the disk, to retrieve the information and transport it to the screen.

To illustrate these concepts, let's set up a hypothetical computer-based application. Imagine you are setting up a system so as to be able to look up magazine articles by subject, author, and title. You allow 35 characters for the title of the article, 30 characters for the name of the author, and 31 characters for the name of the magazine, and 4 characters for the date of publication. You limit the amount of descriptive text to 5 lines of 40 characters each, making that field 200 characters long. As a result of defining these fields, each "electronic card" (record) will have a total of 300 characters.

You communicate this information to your database program, first, by selecting the **define fields**, **create file**, or a similar choice from the menu displayed on the screen and, then, entering the following information:

Field Name	Length	Data Type
TITLE	35	A
AUTHOR	30	A
MAGAZINE-NAME	31	A
DATE-MONTH	2	N
DATE-YEAR	2	N
DESCRIPTION	200	A

This information tells the database program the name you have assigned to each field, the maximum number of characters for each field, and the type of data you will be storing in each field; A stands for **alphabetic data** (A–Z) and N stands for **numeric data**.

By telling the program whether the data in a particular field will be alphabetic or numeric, it will be able to **edit-check** the data that are entered. For example, if you accidentally type in "FE" in the date-month field, the program will reject the information since you've told it that only numbers should appear in that field. Some systems will even beep to alert the data entry operator that she has mistyped something. Some fields, such as those with addresses, may contain both letters and numbers. They contain **alphanumeric data**.

Your Computer Becomes a Filing Robot

You may think that since the work of defining the fields is done, you can begin to use the system. But that's not true. You can't assume that your database system is intelligent. It's only a program that turns your computer into a dumb, but fast, robot filing clerk that takes instructions so literally that it's often maddening.

We've all been brought up on science fiction concepts of humanoid robots that roll along on tractor treads and have movable arms and a head that rotates 360 degrees. Actually, our database management program does all the functions of a robot filing clerk without looking the part. It opens a filing cabinet drawer by accessing a disk. It sorts, files, and organizes information without apparent motion. It types by sending electronic information to the printer instead of moving its robot fingers. But, like a robot, it has no intuitive sense of how to do things right. Therefore, we, as humans, have to instruct it to do many things that we take for granted when we ask a human file clerk to do something.

When you say to a human file clerk, "Please go get Deborah Brecher's record," you are implying more than you may think. The file clerk assumes that the file is organized in alphabetical order, by last name. She also assumes that you are keeping the cards in ascending order, with "A" in the front of the file and the "Z" in the back. You don't need to tell her all this in order to have her bring back your record.

With an electronic filing system, you can't assume anything. Remember, these are generic systems. The same database package that works for survey researchers will work for the property tax collectors, the Department of Motor Vehicles, and telephone directory assistance. Each of these users must tailor the system to suit its particular data needs.

This means that for each user the system must be taught not only a unique record layout, but also the way the file is organized. This allows the Motor Vehicles Department to look up information by license plate number, the property tax collector to look up parcels by address, and directory assistance to look up telephone numbers by last name.

Defining the Way the File Is Organized

After defining your record, you have to do one more bit of tailoring. You have to choose how you want your file organized. In other words, you have to decide which field will determine the order of the records in the file. This also determines how you look up information in the file.

In our example, the human file clerk assumed the membership records are kept in alphabetic order, and by last name. But that assumption isn't always correct. The property tax collector's records are organized by street address. The property owner's name is simply an incidental field of information that gets updated every time the property is sold. What is important to the tax collector is to be able to *look up* information on any given address. That way the current owner can be located if a special assessment is being made on that property, or unpaid taxes can be recorded.

The Motor Vehicles Department organizes its records in still another way—by the license number field. This permits a traffic cop to run a check on outstanding tickets, by giving the computer the license number.

The Key Field

There's a special computer word for the name of the field that organizes the file—it's called the **key field**, or simply the **key**. The key enables you to query, or look up, a particular record from the file. The important thing about the key is that the information in it must be *unique* for each record. For example, you couldn't use a zip code field as the key because so many people have the same zip code that you wouldn't be able to query (look up) one particular record from the file.

The key field is extremely important, and care should be taken in choosing it. Last name/first name, an obvious choice, is not always the best one. In a large file there will probably be more than one person named Elizabeth Smith, for example. Also Elizabeth may sometimes call herself Liz. As humans, we can deal with this. But computers cannot. When we try to look up information on Liz Smith on our computer, we won't find anything. This is because the electronic filing system does queries using the compare ability of the CPU. It compares the key of the record we are looking for to the keys of every record in the file. Only when it gets an *exact* match, does it move the record to the screen. Liz and Elizabeth would never match.

Some companies, organizations, and schools use your social security number as the key. It's a very handy identifier, since you are the only one with your particular number and it's short—only nine digits long. The only problem with using your social security is that in order to look up your record on the computer, they need to know it.

Many large companies create a special field to use as a

key, which is actually a combination of information from various other fields on your record. The key can consist of the first five letters of your last name, followed by the first three numbers of your address, followed by the first three digits of your zip code. You'll often see codes such as this on the mailing labels of magazines you subscribe to. The set of letters and numbers at the top of the label which corresponds in some seemingly strange way to your name and address is your particular key. When you move, most magazines ask you to send in an old mailing label along with your new address so that they can see what your key is. Using your key, they can call up your personal record from out of millions of other records in order to update your address field.

The fact that the key provides each record with a unique identifier also allows duplicates to be discovered. For example, at the dentist's office you may be asked, "Are you a new patient?" If you're in pain and a little dazed, you might misunderstand the question and say, "Yes," when you aren't really new at all. However, when the clerk types your new record into her computer, the system will reject it since it already has a record with the same key.

Once you define the key field, your tailoring work is done and you are ready to start using your electronic filing system.

Data Entry

The first thing you have to do is enter your data; that is, type in your records. To get started you have to tell the program what you want to do. You do this by using the program's menu again. From the menu of possibilities, you indicate that you want to do **data entry**. Every system has its own name for data entry; some call it **update, insert,** or **add records.**

Typing the information on the computer for each record

(electronic card) is no different from typing it on a typewriter. You have one "card" on the screen at a time, and it shows you the names of the fields you have selected. In our example, this would appear on your screen:

TITLE   ....

AUTHOR   ....

MAGAZINE-NAME   ....

DATE-MONTH   ..

DATE-YEAR   ..

DESCRIPTION   ....

                             ....

                             ....

                             ....

                             ....

The dots on the screen remind you how many characters you can put in each field in the same way boxes on paper forms indicate how many characters to write in. As you type, the dots disappear and are replaced, one by one, with the characters you typed. That's all you do! After entering the information on one "card" the program files it and then puts another blank "card" on the screen on which you can enter your next record.

Using Your Database

Now you can use your database system. Keep in mind that everything discussed thus far is only done once—when you first install the system.

Data entry occurs every time you add a new record, for example, every time you catalogue a new magazine article, add a new subscriber, get a new client, or whatever. To look up something, you merely enter the key of the record that you want to see. The desired record is then almost instantaneously displayed on the screen.

Selecting a particular *group of records* is a somewhat slower process. For example, suppose you want to do a mailing to club members whose memberships are up for renewal. In a manual system you'd have to look at every record, one at a time, to see the renewal date. Every record with an expired membership would get "pulled" so that renewal notices could be typed and sent.

Computerized databases work in much the same way. To select (pull out) a group of records that meet a certain criteria, each record has to be moved from the disk to memory (RAM) and compared to see if it meets the selection criteria. This is somewhat time consuming. Depending upon the size of your database, it may take 5 or 10 minutes or, with large files of about 50,000 records, even an hour or more.

Computerized systems are different from manual systems in that the records aren't actually "pulled" from the file. Rather, a copy of each record that meets the selection criteria is written out to disk and stored in a separate file (in computerese this is called a **subfile**). This means that the records meeting your selection criteria can now be used to print up mailing lables, as an address file to merge with a standard form letter ("This is your last chance to renew . . ."), or any other similar use. It also means that you don't have to refile the selected records, as you do in a manual system.

By using database management systems, you can easily generate **special forms** or **reports**. These can be printed out or simply displayed on your screen. For example, you may want to know how many memberships will expire in the next 6 months or to print a list of the names and addresses of all the members whose annual dues are unpaid.

If you wanted the names and addresses to print out using a special format, such as required for mailing labels, a **report definition** would be required. You define a report similarly to the way you define a record, except that for a report definition you specify where on the printout you want a field to appear (instead of defining field sizes as you did on the record definition). For the mailing labels format, you'd have to specify that the city, state, and zip code fields should all print on the same line. It's not necessary to print out all of the fields. For example, the mailing labels will not have fields, such as telephone number, printed on them.

Indexes Save Time

It's easy to see how you create a subfile. But how does a query (lookup) work? How is it possible to pull a particular record instantaneously out of a file that contains thousands of records? If each record in the file had to be searched in order to find the one you wanted, it would be too time consuming.

The answer is that the method of doing instantaneous lookups involves **indexes**. In computer terminology an index is very similar to indexes that you are already familiar with. Think of an index to a book. It is a path into the book. If the index is in alphabetical order by subject, you have a way to look up any subject in the book. You can go right to the information you want (direct access), without having to leaf through page after page.

In a similar way, the database uses an index composed of the key field and the record number. Instead of looking through every record to find a particular one, the database uses the index, which tells it the record number of the record you've requested (instead of page number as in book indexes). Once the record number is known, the disk drive can move its read/write head directly to the correct location on the disk and read the record instantaneously into memory.

The database automatically creates one index using the key field that you selected. But you may also designate other fields to be indexed. All indexes are automatically maintained; that is, when a new record is added to the data file, *all* of the indexes automatically get updated to include it.

This ability to have multiple indexes is one of the wonderful things about computerized databases. It lets you have many ways, based on different criteria, of looking up individual records in your data file. For example, in our magazine article system you can look up things by title, author, publication date, and so on.

You're probably used to manual systems that have multiple indexes. For example, some books have more than one index. In addition to a subject index, they may have an index of peoples names, or a geographical location index. Law books often have case name indexes.

One of the most familiar multiple indexes is the card catalogue file in libraries. The librarian maintains three separate card index systems: subject, author, and title. These card files provide you with three paths into the bookstacks. You are directed right to the book you want, whether you know it by subject, by author, or by title.

If the library's card catalogue were computerized and you wanted to look up this handbook but you only remembered my name, you could go to the terminal and indicate that you wanted to do a query (lookup). The system would then prompt you with something such as WHAT INDEX? and you would answer AUTHOR. A blank index card would then appear on the screen, with the cursor blinking on the field for the author's name. You would simply type in BRECHER, DEBORAH, and up would come the electronic card for *The Women's Computer Literacy Handbook*. If you selected the title index, the cursor would have been blinking on the title field. Then you would type in WOMEN'S COMPUTER LITERACY HANDBOOK.

When using a manual card file you sometimes find yourself leafing through the cards. For example, in the library you may not be sure of the exact title of a book. If you were trying to locate this book, you might open the title card drawer to WOMEN'S and then look through each card until you recognize the name of the book you wanted.

Computer files work the same way. To open the "electronic drawer" to WOMEN, you type in WOMEN'S*. The screen will display the first reference for a book title beginning with that word. If it wasn't my book, you could look through all of the "cards," one by one, simply by hitting the

ENTER or RETURN key to bring the next reference to the screen.

Remember, computer look-ups require the key field to exactly match. Opening the electronic drawer to WOMEN'S was only possible through the use of the special symbol *. In computerese, the * is called a "wild card."

Unlike the librarian who had to type a separate reference card for each of the three indexes (Figure 4-2), as a computer user you can have additional indexes maintained for you automatically by the database. You only have to enter the data once.

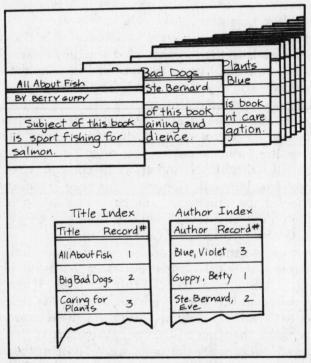

FIGURE 4-2. This computerized card catalogue has two indexes. To look up information by author you use the author index. The database translates the record number in the author index to the physical location of the "3×5 card" on the disk.

Shopping for a Database—Look for Limitations

When choosing a database system for your projects, remember that every computer program has built-in limitations. This is true no matter what database software you buy. The trick is to understand whether or not a particular database program's limitations are going to affect the kind of database you want to set up. The kinds of constraints that are built into database systems include the maximun field length, the maximum number of records, the maximum number of indexes, and the maximum number of fields. How important each of these constraints is depends upon your particular application.

Maximum Field Length

One thing that you have to make sure about when you buy a database management program is that the maximum length of a field will be long enough to fit your data keeping needs. Maximum field size refers to how many characters the largest field can be.

I have a computerized mailing list in which the largest field is the name field, which is defined as 30 characters long. My field size requirements are very different from that of a database system used to maintain magazine abstracts. In a magazine abstract system, the descriptive text would occupy one field. This text would be expected to run to at least several lines of 80 characters each. Therefore, it would be necessary to use a program that allowed a field to consist of at least 400 characters in order to accommodate five lines of text. A program whose field size is limited to 100 characters could not be used for this magazine reference system, whereas it could easily be used for my mailing list without ever feeling the effect of this limitation.

Maximum Number of Records

Each program is also designed to be able to handle a maximum number of individual records. This is of no real concern for a small organization that wants to create a membership database. However, national organizations with 50,000 or more members could not use all the database software on the market since not all of them can accommodate that many records.

Even if the number of records you'll be keeping now is relatively small, it's important to keep in mind that your mailing list or other database system will grow in size over the years. A good rule of thumb is to buy a system that will suit 5 years of growth. The maximum number of records should be able to accommodate your most optimistic projections of the number of clients, members, books, or whatever that you will be keeping track of 5 years from now. This is true even if it means buying a slightly more expensive program now. In the end it will prove to be a greater savings than having to buy a new software program later, not to mention the time and energy it will take to change to an unfamiliar system.

Maximum Number of Indexes

For many database users, the number of possible indexes is not an issue. This is because they will only want to look up things by one field, such as name. For other users, such as the highway patrol, who want to be able to call up records using more than one field, multiple indexes are essential. When you're stopped for speeding, for example, a computer check can be run on you based upon your driver's license number, the car's license plate number, or your name. Again, deciding whether or not you will need database software that allows multiple indexes depends on exactly how you will look up information in your electronic file.

As you might expect, the more capabilities that a program provides, the more the program will cost. Each package has a different set of limitations. Knowing what your application demands is your first step. Looking at the limitations of the packages is your next step. Major brand names that you may want to research include dBASE II or dBASE III, FMS-80, PFS:File, Perfect Filer, DataStar, and R:base.

Uses for Database Systems

Database systems may be the most exciting application for computers. Many services that we take for granted are actually database services. For example, rental car, airplane ticket, and theater ticket reservation systems all use database management software.

Our society is becoming an information society, and the ability to store and retrieve information has formed the basis for many new businesses. One company, for example, keeps track of insurance rates using database software. You tell them what kind of insurance you want, and they tell you which company has the lowest rate. Another company lists apartment vacancies. They don't charge the landlord to list, but they charge the renter to search the system. The rental files have multiple indexes so that listings can be located by criteria such as amount of rent, location, and number of bedrooms, making it easy to search for places that meet a particular need. Roommate services do the same sort of thing. Some employment services keep resumes on file, indexed by job qualifications. In California there is even a high-tech employment agency that maintains its database of available jobs **on-line**. This means that if you have your own computer or terminal and a modem, you can call up their service and search their files from your own home.

It's possible for a computerized database to be maintained so that people could call up a telephone number and get instant referrals to the exact type of services for which they are looking. The services could be referenced by a wide assortment of fields so that the referral operator could select those that you specify and come up with several selections for you from the database. Does this sound like a good idea? It's such a good idea that the telephone company wants to do exactly that. Imagine a computerized yellow pages that you can call up by phone/modem and see on your terminal or monitor. You wouldn't just get the address and phone number, you'd get today's sale prices and an announcement of the specials that will begin next week. Illustrations of products would be included, too. It's such a powerful idea that it may revolutionize the whole concept of advertising. Newpapers are afraid it would put print advertising out of business, so they've been fighting the idea of electronic yellow pages for some time in the courts.

The Darker Side of Database Systems

Naturally, there are drawbacks associated with database systems. First, there's the trend toward fee-for-information service, with pay systems replacing ones that were formerly free. There's also the abuse of privacy that seems inevitable now that it's technologically possible for a microcomputer system to hold a great amount of information on people. The kinds of invasive databases that only large government agencies and a handful of private companies used to be able afford are now less expensive and can be operated by almost anyone.

Paying for Information

The prospect of charging for information that used to be free is an unpleasant one. Many experts predict that the library of the future will be one in which we pay for service. For example, reference books such as bibilographic indexes may simply cease to exist in print form. Instead, they will be part of large computerized databases that require libraries to pay both a monthly subscription fee and a per-use fee. The libraries will have no choice then but to pass on these costs to us, the library users.

In a similar way, apartment rental listings in the newspaper are shrinking, as landlords save themselves the expense of advertising and, instead choose to list for free with a computerized rental service. The cost of the service is passed on to the consumer. Instead of paying 25¢ for a newspaper, apartment rental databases services may charge $35 or more. This doesn't seem so serious, but as information that used to be free gets replaced with fee-for-service, the poorer members of our society will lose even more ground.

Another example of the trend toward paying for information that was once provided free is brought to us by the telephone company. Now that AT&T has been broken up, the phone companies are charging for information. They started with local directory assistance fees and added long distance directory charges. This database, which was once maintained as a free service for customers, is now being turned into a separate profit center, with new "products" being added such as fee for addresses (including zip code) and a reverse directory service whereby you can get a person's name and phone number by providing just a street address.

Now that there's money to be made on this, several competing services are also being organized. With competition, these new services will probably offer additional information, such as people's work numbers and birthdates. Frankly, all this makes me nervous. For example, the reverse directory service sounds like a handy tool, but it has dangerous potential. A burglar, for example, could get the phone number for a particular address and call up to see if it is vacant. Although such information is currently available, it takes some effort. These new services will make the information handier and more immediately accessible.

Losing Our Privacy

A more chilling problem comes with database applications such as dossier services. There are already companies that for a fee will amass a dossier on you. Using their computers with modems, they search a lot of public records that are maintained on-line. When they access the census files, which contain financial information by census tract, they can find average yearly income. Car registration, home ownership records, and other such innocuous pieces of information can be combined with your credit rating to provide a complete analysis of who you are.

A frightening and little discussed assault on individual privacy is the fact that many organizations, businesses, and magazines sell their mailing lists without permission of their members, subscribers, or patrons. Dossier companies and even the government can buy up these lists to see what names are on them. Think about it—the type of magazines you read reveals more about your political affiliations than your voter registration does. And the stores you shop at say something about your lifestyle. In fact, the IRS decided to use this approach to look for people who have avoided paying taxes. (They haven't yet used this approach to look for people who underreport their incomes.) They bought mailing lists of chic and expensive magazines and used their computer to cross-check the names to make sure that these supposedly affluent people on the lists filed income tax returns.

The Orwellian "Big Brother society" is too close for comfort. The problem is that there are no laws that regulate the buying and selling of information. There are a few regulated industries, such as banking (thank goodness), but on the whole, anyone can buy and sell information on you. There is the Freedom of Information Act, but that just gives you the right to find out what information a particular government agency is keeping about you. It doesn't say anything about regulating the type of information a private company, or even the government, can keep about you or whom they can sell it to.

Most people have their own information horror stories about something that happened to them. Incorrect information that has somehow crept into a file can come back to haunt you, whether it's a bad credit rating, a late payment to a department store, or an unpaid traffic violation that you actually paid. It's because of the problems associated with incorrect information that credit bureaus are regulated. In most states, you are entitled to find out about the information in your credit file anytime you want to. If you've been turned down for credit, the credit bureau has to send you a copy of your file for free; otherwise you have to pay for it. But what about all the other files that are being kept on you?

A Modest Proposal

My personal recommendation is to mandate that any company or agency that keeps information on you be required to send you a copy of the contents of your file once a year. It's already on their computer; therefore, printing the information and sending it to you is relatively easy. After all, banks do exactly that—they regularly send a statement of your account so you can corroborate that the bank records are correct. Why should any private company that makes money by keeping information about you have to do less? And, most importantly, the burden should not be shifted to the public so that you would have to take the time and effort to request your file from each company that maintained information on you. So few people would actually avail themselves of this right (how would they even know that information *was* being kept on them) that inaccurate recordkeeping and reporting would go unchecked for the most part.

Ethics of Databases—
A Positive (Feminist) Model

Now that you have been warned of the worst consequences of implementing computerized databases, let's look at how a database can be designed that incorporates ethical considerations and produces exciting results.

One of the positive effects of databases is that they create new possibilties for networking. In 1981 I noticed that all over the country there was an abundance of women's services and businesses that had been failing simply because they didn't have any way to let people know that they existed. They were providing desperately needed services, such as battered women's shelters and rape crisis intervention, doing advocacy work for women's special interest groups, such as tradeswomen, older women, and minority women, and contributing to the advancement of women's culture, such as small press publishers of women's titles and producers of women's art, music, theater, and film. Unfortunately, even the most successful of these ventures did not have the money to advertise in the press or on radio, to say nothing of television. And as a result, their efforts went largely unnoticed by many of the women they were created to serve. At the same time, women who wanted to locate these services had no way to find them.

To meet this need, I started a computerized database called the National Women's Mailing List. The mailing list served as a networking function and as a demonstration project to show that computerized databases containing information about people could be implemented according to feminist (ethical) principles.

I call these principles feminist not because I think they are necessarily limited to women, but rather because they have been articulated most persuasively by women who have addressed these issues as feminsts (for example, Judy Smith of the Women and Appropriate Technology Network and Corky Bush of the American Association of University Women). These principles include

- Respecting the rights of the individual
- Having the participants control the system
- Letting the users participate in the design of the system
- Recognizing in advance the consequences of the inevitable failure of the system: It *will* break!

The National Women's Mailing List

To see what these principles mean in practice, let's look at the National Women's Mailing List in detail. Most computer-based mailing lists are created by acquiring other lists of individuals and combining them into one database, without the knowledge or consent of the people whose names are included. In contrast, our mailing list was voluntary. To be on it, you signed up on a special registration form, either as an individual or as a women's organization.

The registration form contains numerous fields, allowing the participants to identify themselves in terms of their age, occupation, cultural identity, and parental status or to describe the purposes of their organizations. The individual also indicates which subjects she wants to receive mail about, choosing from an exhaustive list of women's interests. Each of these interest areas represents a separate field within the database system. As the forms come in, each one is entered into the file as a new record.

Using the selection capabilities of the database management system, we can build custom-designed lists for a wide variety of women's organizations and interests simply by choosing which fields to search and compare. For example, we can have the computer read each record from our master file (the database) of 60,000 names and extract all those that indicate an interest in women's literature. Feminist publishers can then send announcements about new books to these people. Several fields could also be combined. We could select people interested in women's literature who also live in Chicago so they could receive advertisements about book-signing parties in the Chicago women's bookstore. If the book was about older women's issues, we could find people interested in women's literature who live in Chicago and who are also over 50 years old. If the book was a text-book, we could add educators to our list of criteria. As you can see, a single database can serve a variety of needs, depending on the number of ways it can be accessed.

Not only is the National Women's Mailing List database voluntary, but it respects the individual choices that the participants have made. Woman who have little in common can be part of the same database network, with the assurance that they will only receive mail about the subjects that they themselves have chosen. A woman who signs up and indicates that she is only interested in women's health issues will never be selected for a mailing list that is being created for people interested in women's sports. On her electronic record, the data field for sports is blank so there's no chance that her name will get onto the list that was ordered for the Conference on Women in Athletics. The computer, being always faithful to its instructions, will look at the field for sports on her record; since it will not compare favorably with the criteria it's been given, it will move on without including that record in its selection subfile. Our computer absolutely respects the choices that individuals have made about how they want to participate in the system, rather than the system imposing its choices on the individuals.

THE REGISTRATION FORM

Since the usefulness of a database depends entirely on what fields of information have been included, we realized it was important to have a registration form that was as exhaustive as possible, listing all the significant interest areas people could choose to get mail about. Therefore, we started by doing our own survey, but recognized that there would be areas that we would overlook. To discover what they were, we included many blank lines labeled "other" at the end of each category of interests and monitored the responses carefully. In this way, over 20 additional categories have been added to our initial design. By allowing the people involved in the system to have input into its design, the number of possible uses for the database was increased to everyone's benefit.

Using the computer, it is relatively easy to change the layout of the form to include new data fields. Of course, the old members will have blanks in the new fields, but from the time of the redesign forward, the new information will be collected for future use. This approach rejects the idea of the omniscient system designer and recognizes that every good system should be built to anticipate future changes.

We knew that there was no way to know every issue of interest to women. Our role became that of information managers, or database librarians, modifying the record definition on the computer in response to patterns of requests. This approach has the added benefit of working with a "living" system, able to respond to changing social conditions. For example, when the system was designed, the nuclear freeze was not an issue. When repeated requests for networking around that issue were received, it was added to the existing interest areas.

SYSTEM DESIGN AND FAILURE

Finally, the system design takes into consideration the possibility of failure. The realistic (non-arrogant) approach is to assume that the system *will fail*, and you should plan accordingly. The odds that the system will fail tomorrow are extremely low; but the odds that the system will fail someday are close to 100 percent. Therefore, good backup and recovery procedures are built-in.

Just as importantly, the system does not collect information that would be harmful if it somehow got "leaked" into the wrong hands. This is a very important design criteria.

Even though the computer system is secure–through the use of secret passwords for access to the computer and encryption of the data–it is still important to remember that every system, no matter how safe it appears, can be broken into.

Taking this into consideration, we do not collect information on what organizations a woman already belongs to. Many women's groups have asked us to collect that information so that when they used our lists they would mail information only to those people who were not already part of their membership. But a list of organizational memberships begins to approach building a dossier file, which is exactly what we want to avoid. Even though we limit our service to providing mailing labels containing only name and address and do not disclose the entirety of a person's registration form, we do not ask our participants to disclose any information about themselves that might someday prove harmful, just in case our records got leaked.

I've gone into great detail about system design because these design criteria are simply not discussed in most computer reference manuals. Like many professional texts, the technology of *how* to do something is separated from the moral questions about its *appropriateness*. If the idea of a built-in assumption of system failure were more widely accepted, for example, no one would be willing to take the risks associated with nuclear power. But it is important to remember that our own common sense tells us that everything breaks sooner or later. The real question is, are we prepared to live with the consequences of a system's failure?

CHAPTER **5**

Software Applications: Spreadsheets

The third type of software available is referred to generically as **spreadsheet** software. Basically, spreadsheets allow you to set out figures in columns and lines, just as it's done in bookkeeping and on ledger sheets. The most common applications of spreadsheets are budgeting and financial planning. Like database software, spreadsheets are generic. What this means is that initially the spreadsheet displays a blank screen as shown in Figure 5-1. Before entering data, you must tailor the spreadsheet by defining the **rows** and **columns** to fit your needs.

Figure 5-2 (see next page) shows a typical table of information you may include on a spreadsheet. Lines for each of the months and for the yearly total have been defined.

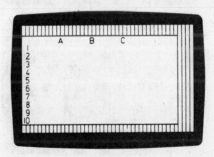

Spreadsheet

FIGURE 5-1. When you purchase a spreadsheet, it must be tailored before you can use it. Before labeling, the rows and columns are referred to by numbers and letters.

151

Column headings for each major expense category, in this case rent, utilities, supplies, and salaries, have been specified. Actual data is entered into each cell—the intersection of a row and column—by moving the cursor to the appropriate cell.

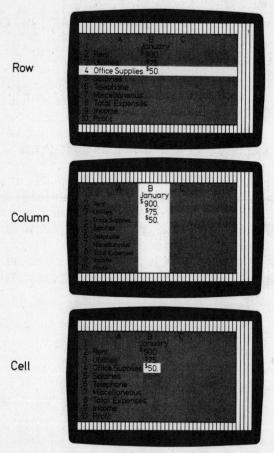

Row

Column

Cell

FIGURE 5-2. This is a typical table of information you may include on a spreadsheet. Tailoring a spreadsheet allows you to display your own labels for the rows and columns, such as those in this table. You enter data by moving the cursor to the appropriate cells.

A spreadsheet program turns your screen into a grid of cells. Since each cell is identified, it's easy to refer to them. Different programs have their own ways of referring to specific cells, but one common way is to identify them by a letter and a number. The letters correspond to the columns and appear along the top of the spreadsheet, and the numbers correspond to the lines and appear along the sides of the spreadsheet. As Figure 5-2 shows, cell B4 refers to the amount spent on supplies in January.

A spreadsheet allows you to keep data in columns and rows, but this could be done using a word processing program as well, with just a little less efficiency. Why, then, are spreadsheets so popular?

Why Use Spreadsheets?

We've all had the experience of adding a column of figures and then realizing that we left out a number or put the wrong one in and had to add the whole column again. Spreadsheet software solves this problem. The spreadsheet can be instructed to add all the numbers in a given column, (or in any cells that you tell it to), and to place its result in another cell, such as the one you label "total." This means that you can add, delete, and change any of the numbers you are using without having to go back and recalculate everything.

If this were the only claim to fame for spreadsheets you could use a calculator and save a lot of money. It's not. The real power of the spreadsheet programs is its ability to define columns or rows in relationship to each other. For example, in the spreadsheet for monthly expenses that we described in Figure 5-2, we could include last year's spreadsheet plus this year's projected budget in adjoining columns. Then we could have the program compare our telephone expenses for the current month to those for the same month last year and for year to date and give us a percentage increase or decrease. This percentage figure could be automatically placed in the cell next to this month's telephone expenses. In the cell next to that, we could put the result of the ratio between the year-to-date's telephone expenses and the total amount budgeted for the year to show what percentage of the telephone budget we already spent.

The spreadsheet can be taught to do just about any arithmetic operation with your data. You could tell it to take the data in each of the month's "total" cells, add each of these together, divide by 12, and put the answer in another cell labeled "Average monthly expenses for the year."

Because spreadsheets make it so easy to adjust figures, people often use them to do projections based upon a variety of assumptions. By going into the data and changing figures to see "what if . . . ," the result can be instantly calculated. This feature allows for quite sophisticated analysis.

Spreadsheets Work on Any Table

At a certain point, people started to realize that they could use spreadsheet software for applications other than budgets. For example, a factory could do product pricing using a spreadsheet. Suppose a bakery was trying to figure out what to charge for brownies. Each of the ingredients could be entered by their cost per weight (that is, cost per pound, gallon, etc.). Then the cost of a pound of sugar, a gallon of milk, a quart of nuts, and so one, could be entered in specific cells. Other cells could contain per-hour labor costs for the bakers and overhead expenses. The pricing formula for brownies could then be defined in terms of weight of each ingredient, hours of bakers' time, plus the percentage of plant overhead per dozen brownies, and a cost figure for brownies could be calculated by the spreadsheet. The best thing about this system is that if the price of anything goes up or down, the spreadsheet will automatically calculate the result so that the price can be raised or lowered to reflect the product's actual cost.

In a similar way, building contractors, carpenters, and other tradespeople use spreadsheets for preparing bids on jobs. Just as in the bakery example, the component parts are defined to the system. But instead of the product cost of a brownie, the bid on a particular job is calculated and adjusted depending on how much subcontractor's or employee's time would be required at the particular pay rate plus the amount of materials needed and their costs. Every time the price of a certain commodity goes up, it is possible to rerun the spreadsheet program, adjusting only the figure that changes—the price of that commodity in this case—and the entire pricing structure would be automatically adjusted. For the bakery, new prices for each type of baked good would be calculated when the price of sugar went up. For the builder, a new figure for the cost of a job would be generated every time the client added on something extra or the price of lumber went up.

Other complicated what-if questions naturally lend themselves to this type of analysis. The decision on whether to lease or purchase equipment is one common use of spreadsheet software; here the answer depends on things such as tax brackets, tax credits, interest rates for financing purchases, and monthly costs. Once the method of analysis is defined to the system, you can vary the outcome by plugging in different data. A friend of mine is a labor negotiator and uses his spreadsheet program to see the results of certain bargaining positions. The tradeoffs between benefits and salary can be analyzed. The employer or a union representative can ask the spreadsheet software to weigh the relative merits of a higher salary versus more health benefits or holidays.

In all these examples, the spreadsheet software must first be "taught" the specifics of the particular analysis, whether it is the formula for pricing brownies, the dollar value of a health insurance policy, or the calculation of year-end totals. This process of tailoring is sometimes called **modeling**, since a model of the company or product is created through the process of defining all the variables that go into it.

Formulas Have Been Created for You

In various businesses and industries people have already figured out the way to arrange their spreadsheet grids, complete with all the formulas necessary for making sophisticated analyses of the data. Some spreadsheet software come with these grids. But more often you will need to purchase the special layouts to be used with your brand of software.

These layouts are also known as **templates**. They act like transparent overlays on which you enter your data. For example, you can buy templates that adjust your spreadsheet to resemble an income tax 1040 form or other IRS schedules or forms. Other templates tailor your spreadsheet into a profit-and-loss statement program.

Differences among Spreadsheet Programs

With spreadsheet programs, as with all software, you are buying limitations. These include the maximum number of cells, the maximum cell width, and the flexibility of display formats. The ability of the spreadsheet program to display data graphically as pie charts or graphs may be essential to some users and an unnecessary feature to others. If you shop carefully you can obtain only those features you need and avoid paying for features of no interest.

Maximum Number of Cells

All spreadsheet programs limit the *number of rows and columns* allowed. The total number of cells ranges from 100 to 1,000 or more. For many users, an adequate number of rows are available for most basic accounting and budgeting uses.

However, if you have many entries, such as for all the items of inventory in your stock, you must make sure that the spreadsheet program can accommodate them all. This is equally true with the width of the spreadsheet: Does it have enough columns to fit your record-keeping needs?

Cell Size

The *size of the largest cell* may be of concern to you as well. Almost all spreadsheet programs allow you to set the width of each of your columns, but most programs have a maximum cell width that's allowed. Setting the width is important because not every column will hold the same sized data. Consider this example. If you were a demographer, you might have a spreadsheet on which you kept track of the populations of various countries. The column would have to hold a maximum of 13 characters to accommodate China's 1,000,000,000 population figure. However, the next column might be for average family size, which would only require a width of 3 characters to hold a figure such as 3.5. By specially defining each column's width, you won't have a lot of empty space and you'll be able to keep more information visible on your screen at one time.

In most spreadsheet software, the columns are preset to hold 8 or 9 characters, and in the better systems, you can have widths as large as 80 characters. In the less expensive spreadsheet programs, however, you cannot exceed the original width of 8 or 9 characters in a cell.

If you wanted to use a spreadsheet package to keep track of your payroll, with separate columns for the employee's name, social security number, hourly rate, gross pay, employer's contribution of social security withholding, employee's contribution of social security withholding, federal income tax, state income tax, net pay, and totals along the bottom line, you would need columns to be considerably larger than 9 characters. There wouldn't be enough room to hold each employee's name (for example, Donna Williamson) unless the name column was at least 20 characters wide.

Seeing the Whole Spreadsheet

One of the more uncomfortable things about electronic spreadsheets is the difficulty of seeing a large spreadsheet on the screen. If you have many columns or rows, there is no way to see more than part of your spreadsheet on an 80-column-wide monitor at one time. There are several approaches to this problem.

Some hardware manufacturers, such as Digital Equipment Corporation (DEC), sell 132-character-wide monitors. This helps somewhat, but frequently the number of columns and the widths even extend beyond 132 characters. You can always move your cursor around to various parts of your spreadsheet, but this may not allow you to see more than just the part of the spreadsheet on which you're working.

This problem is particularly acute if you are trying to see data in a column (or row) that is far away from its heading. For example, on a payroll information spreadsheet you might want to look up the net pay for employee Williamson, but the first column on the far left has her name, whereas the last column on the far right has her net pay. If the spreadsheet had many columns in between, the two fields in which you're interested couldn't be displayed on the screen simultaneously. You could see the far right or the far left, but you could not see it all at once. If you were looking at the net pay column on the spreadsheet's far right, it would be difficult to know which cell was the one for Williamson.

To deal with this problem, there are **windows** in most good spreadsheet programs. This allows you to split the spreadsheet at a given point and bring up another section to sit beside it on the screen. In our example, you could split the spreadsheet after the column for employee name and then bring up the last columns from the right side of the spreadsheet to appear next to it. In this way you won't be confused about whether you're looking at the correct cell, since the name Donna Williamson will now appear right next to the column for net pay.

Graphics

One of the biggest differences among spreadsheet programs is their ability to do **graphics**. Many people have a hard time fully comprehending the relative weights of numbers. Graphic presentations such as bar graphs or pie charts are an excellent visual way to present this information. See the

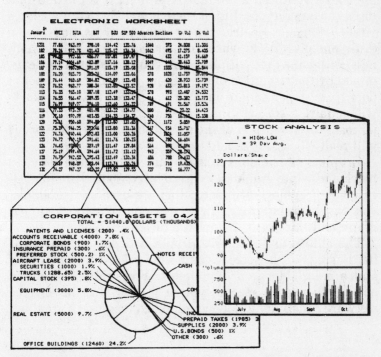

FIGURE 5-3. The figures in a spreadsheet can be graphically displayed as bar charts, pie charts, or line drawings. (Courtesy of Micro Software Systems.)

comparison. Figure 5-3 shows you some of the types of displays that can be provided by graphic display software.

Like word processing software, many spreadsheet programs come in **modules**. In these cases, the graphic capabilities may be sold as a separate module. Other companies sell spreadsheet packages that include graphic capablities as part of the base price. Still other companies sell **stand-alone** graphic display programs to work with the data you've created using your spreadsheet.

In any case, the quality of your output and, in fact, your ability to get graphic output at all, will also depend on your peripherals. For example, in order to be able to print out a pie chart, you need a dot matrix printer. It is the dots that draw the circular pie. Just as there is no way to draw a large circle using an electric typewriter, there is no way to draw one on a daisy wheel printer. Similarly, the "roundness" of the pie that's displayed on your screen will be determined by the resolution of the screen; that is, the number of dots (pixels) per square inch. Many spreadsheet packages will require a special graphics controller board in order to increase the resolution of the screen so that the circle or pie will really look as it should.

Graphics can be done in black and white and in color. In black and white graphics, items to be compared are shaded. In color graphics, different colors are used. For example, you can have last year's figures in one color on your bar graph and this year's figures in a different color.

The ability to get colored graphs is dependent upon the hardware. For the screen, you need a color controller board and a color television or monitor. For printed output in color, you need a special type of printer. Just having the software capability is not enough when it comes to graphics.

One Program Works Many Wonders

The three types of software described in Chapters 3, 4, and 5—word processing, database, and spreadsheet—provide most of the data processing power needed by small businesses. But there may be problems associated with using differing brands of software for different tasks. The big question is this: "Will all the *separate* programs work with one data file?"

Historically, the answer to this question was often no. That meant that a letter prepared using a word processing program could not be merged with a name-and-address file maintained with the database program. Instead, the names and addresses had to be entered twice—once into the database program and once into the word processing program's merging file. Similar problems occurred when reports were prepared. Often the tables that were generated from the spreadsheet couldn't be combined in the word-processed report. Either the table had to be retyped using the word processing program, or the old method of physically cutting, pasting, and photocopying had to be used. This led to the messy situations illustrated in Figure 5-4 (see next page).

Integration

Software designers, recognizing this problem, agreed upon data standards. This led to a new generation of software called **integrated** software. This meant that different programs could work with one data file. For example, the numbers in a database file could be displayed in a spreadsheet format. Similarly, the cut-and-paste function of word processing could be used to move a spreadsheet's table into the document being word processed. A word-processed letter could be merged with the client information maintained in a database.

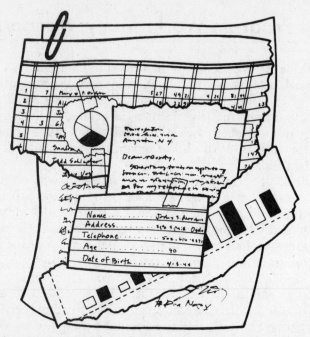

FIGURE 5-4. Before integration, you had to print out the material and cut, paste, and photocopy it into one form. This could be very messy as illustrated here. Integration means you can combine the data generated by two or more programs. For example, the text of a letter might be merged with the name and address in a database file; the figures and graphs prepared by a spreadsheet could also be merged into the letter.

Advertisements for software usually describe the extent of integration that their programs provide by saying things such as "This program integrates with WordStar" or "This program integrates with SuperCalc." That way you can choose the programs that work together (with the same data). For example, in choosing FMS (a database program), it's good to know that its data can be used by both WordStar and SuperCalc.

Although this situation sounds ideal, there were still problems. One of the most annoying problems was the need for a user to learn different commands for each program. No common conventions were established for even the most elementary functions such as cursor movement. Even more bothersome was the need to leave one program before loading the other program into memory (this loading process is described in more detail in Chapter 6).

The newest integrated software solves these problems. The newest approach is what I call "the one big program." These programs combine the functions of word processing software, database software, and spreadsheet software into one huge program. Usually a fourth type of software is included—the communications program needed to transmit information to a computer at another location.

This approach means only one set of commands needs to be learned. For example, the cursor commands will be the same no matter which function is used. There's also no need to exit one program in order to load another. Instead, a menu containing all the possible functions is displayed.

There are drawbacks associated with this approach, however. The software companies that provide the newest integrated all-in-one programs are the same companies that sell single-purpose software. As you might expect, a specific company's all-in-one software does the function of its particular single-purpose program best. For example, Lotus sells the most popular spreadsheet, Lotus 1-2-3. It should be no surprise that their integrated program, Symphony, has an excellent spreadsheet. Symphony's other functions are somewhat weaker.

Another problem of this one-big-program approach is the need for huge amounts of memory—often 384K or even 512K of RAM. Because of this, these huge integrated programs can only be used on certain brands of computers. Figuring out which programs will work on which machines depends upon other constraints, as well as memory requirements, and is the subject of our next chapter.

CHAPTER 6

Putting It All Together

In chapter 1, we discussed what computers have in common. In this chapter, we'll look at some of their differences. For example, you learned that when you buy a computer you're buying a CPU. Here you'll learn what the differences are among the various brands of CPUs.

I've called this chapter "Putting It All Together" to point out that there is an important interrelationship between hardware and software. This means that in examining the differences among CPUs, the big question is how do those differences affect which software will work on your computer. In other words, what determines **software compatibility**, or whether a particular program will work on a particular computer.

Answering these questions requires an understanding of the relationship between the CPU and the **computer languages** that are used to write the programs. Keep in mind that you, as a user, won't need to learn a computer language nor will you have to write your own programs. But if you want to understand enough about computers to buy one or to shop for specific programs, you do need to understand what a computer language is. This chapter will help demystify the role of computer languages and show you how they fit into the scheme of a computer system.

The CPU—An Advanced Lesson

Most of the differences among computers have to do with
the design of the CPU. As you know, the CPU works very
fast—it can perform hundreds of thousands of instructions in 1 second. The speed at which it works (its cycle
time) is possible because there are no moving parts.
Rather, it uses electricity, which moves at the speed of
light, or 186,000 miles a second. But how does it work?
What's inside the CPU? Once you know that, you'll really
understand the differences among competing brands of
computers.

It will be a lot easier to understand a CPU if you remember that one of its major functions is to act like an adding
machine. In the olden days (before 1960) there were no
pocket calculators. Instead, there were mechanical adding machines that used gears driven by electric wiring. If
you took the back off of an adding machine, you saw the
same sort of wire "spaghetti" that's inside telephones. The
design of the wiring caused the gears to turn in the special
pattern that made the adding machine work correctly.
But all that wiring made adding machines bulky and expensive.

CPUs work in much the same way as adding machines;
however, like RAM they too are made of silicon. Inside the
CPU there are a few high-speed RAM locations that are
called **registers**. Essentially, these registers have replaced the
old-fashioned adding machine gears. Instead of wires, the
registers are connected to each other by electricity flowing
along silicon pathways that have been permanently etched
into the chip. That's why all the gears and wires of the old
adding machine can be replaced by a silicon chip the size of
your fingernail; see Figure 6-1 (see next page).

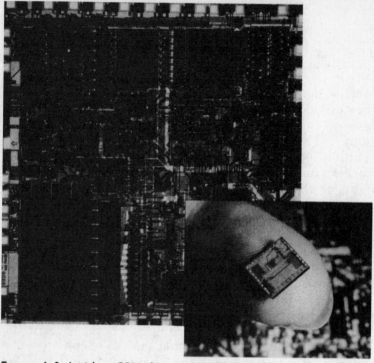

FIGURE 6-1. Inside a CPU chip, circuitry is permanently etched into the silicon. This electron microscope picture of a CPU allows you to see these minature circuits. (Courtesy of Intel Corporation.)

CPUs Are Unique

Computer engineers are constantly competing to design new CPUs that work faster than previous ones. Each new design results in a CPU that will compare or add in less time. Each of these CPUs has its own name. As you might expect, the names for these unique designs aren't very friendly. Computer designers like numbers; therefore there are the Intel chip 8080, the Zilog chip Z80, the Motorola chip 68000, and so on.

Each CPU chip is designed so that a certain pattern of electrical impulses, a binary code, causes particular circuits to be activated. There is no standardization; each brand of chip has its own unique pattern to activate its operations such as add and compare. For example, the binary code you use to get a Zilog Z80 chip to go into its add mode wouldn't get any other brand of CPU to add. Remember that binary codes are used to represent letters, numbers, and special characters; therefore, to distinguish the binary codes that are used to activate functions (e.g. add and compare), they are given the name **function codes**.

We see function codes working when we use common selector switches such as the ones on photocopy machines. If you want to get letter-size copies, you push the button for 8×10 paper. If you want copies on longer paper, you hit the legal-size button. Each of these buttons activates a particular set of circuitry in the copier. That's all that happens when your CPU receives a particular function code; it's like pushing a button that activates a specific circuit. If the CPU receives the function code to add, it will add; if it receives the function code to compare, it will compare; and so on.

Machine Language

One way to write computer programs is to give the CPU function codes to perform a sequence of commands. As the CPU receives a particular function code, the proper circuit is activated and the task you've requested is performed. Because of this I think of the function codes as "CPU language" or "CPU talk." The real name for this "language" is **machine language**.

It's fairly difficult to write programs in machine language. Each brand of chip has a unique set of very specific function codes. Since there's no standardization, a machine language programmer has to learn a new vocabulary for each type of chip for which she wants to write a program.

A greater problem is that there are no simple instructions for common operations such as multiply and divide.[11] Thus, if you want the computer to multiply two values, you can't give it one instruction, such as "multiply 23 times 4." Instead, you have to go painstakingly through each step that constitutes multiplication, as we did in our multiplication program in Chapter 1. Each of these steps is called a **line of instruction** and, as you can well imagine, the seemingly simple process of multiplying two values ends up being many lines of instructions in a program written in machine language.

Intermediate Computer Languages

Because it's very time consuming to write programs in machine language, intermediate languages have been created. Why should each programmer who writes an application program (such as a spreadsheet program) have to keep reinventing the wheel by specifying the same long list of instructions every time she wants the CPU to do a commonplace operation such as multiplication?

Remember our multiplication program in Chapter 1? We had to add the first number, add 1 to a counter, compare the counter to the multiplier, and, depending upon the result of the comparison, either go back to the first instruction and add again or find out we were finished. That series of instructions could have been saved and given a name, perhaps MULT. Then every time a programmer said MULT, our multiply program would automatically get used.

Similarly, other intermediate programs could be written for common operations; there could be one for divide, another for square root, and so on. In fact, that's just what software designers have done. They have written many intermediate functions, saved them, and given them particular names. The names assigned to each function form the "words" that make up computer languages.

The idea of a program to write programs probably seems confusing. Therefore, let's go back to our cooking analogy to clarify this concept. In cooking there is a special vocabulary used to write recipes. For example, in cookbooks you'll often see the verb "sauté." You know that sauté means take out a frying pan, melt butter, and cook the ingredient lightly in it. All of those instructions were conveyed in a single word; when you read that one word, you knew each and every operation to perform.

Similarly, a computer language is just a set of verbs. When a programmer writes an instruction in a computer language that says, for example, "MULT 23, 4," the CPU will perform all of the adds and compares that are necessary to multiply 23 by 4. The computer language automatically converts the instruction that the programmer wrote into the small tedious steps that the CPU must follow. Just as you translated the word "sauté" into its elementary steps when you cooked, each computer language "verb" gets translated into its elemental instructions, which then get performed by the CPU.

Special-Purpose Computer Language

Instead of there being one huge computer language with many verbs, there are a variety of computer languages, each having its own fairly small set of verbs. This approach makes it easier for a programmer to learn the rules of a particular language.

Typically, a computer language is designed for a purpose—to write instructions about a certain subject. This is the way human languages work, too. Just as cooking has its own language, so do sewing, gardening, football and so on. If you didn't know cooking language and wanted to communicate a recipe, you wouldn't be able to say "sauté," or "parboil," or "baste" but would have to painstakingly describe how to perform each of these instructions.

Let's consider the Eskimos, who have a highly specialized language in which there are many words for snow. There is a special word for wet snow, another for dry snow, and many other words for a variety of other snow conditions. Eskimos have created a language with precise definitions for numerous snow conditions because snow is an important aspect of their lives. If you wanted to be able to talk about snow, briefly yet precisely, their language would be the best one to choose.

Similarly, a computer programmer will write a program using the language that has the best vocabulary for that purpose. A subject of many programs is dollars and cents. When you deal with dollars, it's very handy to use a vocabulary that lets you have a word for an instruction such as "line up the decimal points." A verb for this instruction would certainly make it easier to write programs that print out bank statements with the dollars and cents arranged in columns, regardless of the size of the dollar amount.[12]

In fact, there is a language that makes it easy to do that kind of function; its name is COBOL. Most programs for the banking industry are written in COBOL, which stands for *CO*mmon *B*usiness *O*riented *L*anguage. COBOL is simply a set of verbs that are actually small programs written in CPU language. COBOL was created in the early 1960s, by a woman who worked for the Navy, Grace Hopper. She did such a good job of choosing which verbs would be needed by business applications that her original set of verbs is still used today.

In scientific applications it's usually necessary to do a long series of computations before reaching the final answer. For the space program, for example, complicated calculations were essential to figure out the amount of fuel necessary to get to the moon. The result of such calculations is usually a one-figure answer. Therefore, a verb that said "line up the decimal points" wouldn't be required. What was really needed was a language that made it easy to do special calculations such as finding the square root of a number and geometric functions such as tangent, sine, and cosine. That type of verb set is the computer language called FORTRAN, short for *FOR*mula *TRAN*slator.

Another language, called BASIC, was created by two teachers at Dartmouth College. The teachers wanted a set of verbs that would be very easy for beginning students to learn. They designed BASIC to have a simple vocabulary with just enough detail so students could learn how computers worked. It does not have the specialty verbs that other special-purpose computer languages do. In this way, students could learn how to program computers without having to learn a large vocabulary.

Because it's so easy to make up a computer language, people keep inventing new ones. A person who knows machine language can create a computer language just by writing a number of small programs, each describing the instructions to accomplish some common task. Those programs become the vocabulary that other programmers use to write application programs.

Your Need for Computer Languages

Why should you care about computer languages? One reason for you to understand computer languages is to understand what your children are talking about. Like the Dartmouth College students of the 1970s, many elementary school children of the 1980s are being taught BASIC. Some schools are switching over to the language LOGO, which is designed to make graphics easy. LOGO uses the friendly image of a turtle on the screen. Children learn how to make the turtle turn and move by using simple programming commands. But even if you don't have children, it's good to understand what these computer concepts mean, since this machine is becoming a greater part of modern life.

Another reason to understand about computer languages is so you don't pay for computer language programs you don't need. Many people are sold the BASIC language when they buy a computer. But why should you pay for BASIC if you're not planning to do your own programming?

Many people make the assumption that they need to own language programs if they are going to use computers. They think they need to find out the language in which their prepackaged software was written and then purchase that language program in order for their software to work. *That is not true!*

The reason you don't need to own language programs is that although prepackaged programs for business applications (application software) are written in a particular language, such as COBOL, by the time you buy programs they are no longer in their original language. Even though your application program was originally written in COBOL, it was then translated into its elementary machine language instructions, which is what's recorded on the disk you buy.

Translating Computer Language to Machine Language

After a program is written in human-like intermediate language instructions, it gets converted into the CPU's machine language and stored on disk. To understand the conversion process, imagine that the CPU speaks Italian; that is, think of Italian as our machine language. If you could speak Italian, you could write instructions that the CPU could understand. But you don't speak Italian; you speak English. In our example, English represents the human-like intermediate computer languages that you write programs in, such as COBOL, BASIC, or FORTRAN. The problem of translating the language you speak (English) into what the CPU speaks (Italian) is just what you would encounter if you (as an English-speaking person) were asked to give a lecture in Italy.

There are two possible ways to deal with this situation. First, you could bring an Italian translator with you to Italy. In that case every sentence you said in English would be converted into Italian by your translator. Second, you could go to a translator here in the United States and have her translate your speech into a cassette recorder. Then you could go to Italy with your prerecorded Italian speech on a cassette tape and just play the tape.

Let's look at the consequences of these two scenarios. In

the first case, with your translator along, you could interact freely with the Italians—ask directions, order food, and have conversations. Since your translator was along, you could give your speech and ad lib as much as you wanted. If, as in the second case, you just had your tape along, you would only be able to play the tape of your speech. You couldn't alter it or comment further about the subject.

SIMULTANEOUS TRANSLATION

Let's apply these same concepts to computers. One possible approach is to buy a program in a language you speak (we'll imagine it's BASIC). However, the CPU doesn't understand instructions in BASIC; it speaks only machine language. In order to use your program, you will also need to own a copy of the BASIC language program, which you will have to load into memory first. As you use your program, each instruction in it will be simultaneously translated by BASIC into machine language so that the CPU can understand and follow it. When you purchase programs in their human-like language (BASIC, COBOL, FORTRAN, PASCAL, and so on), it's referred to generically as buying the **source code** (Figure 6-2 on next page).

When you buy a program in source code format, you have the same type of flexibility you had when you went to Italy with your translator: You can change things. You can take a program that someone else spent years writing and debugging and figure out how to add instructions to make it do more functions. You can decide to market your altered program under your name and thereby compete with the original software developer.

As you might imagine, most software companies don't like to supply you with programs in their source code version for exactly this reason. Another reason that the software manufacturers don't want you to have the source code is that you could modify the program and inadvertently introduce an error. You might call up the company and complain about the product and ask for help. They might think something

FIGURE 6-2. A program that was written in an intermediate computer language must be delivered to the CPU in the machine language the CPU understands. This can be done by operating the program in its intermediate language simultaneously with the intermediate language program itself. This is like bringing your Italian translator with you to Italy. Most often, however, programs are sold to consumers in object code, which means the program or the disk has already been translated (compiled) into machine language.

was wrong with the original program and waste a lot of time and effort trying to fix what was never wrong.

COMPILING MEANS NO SIMULTANEOUS TRANSLATION

Another approach is to provide the software on a disk that contains the instructions already translated into machine language (Italian). Then you will not need to own BASIC even if its the language in which the software was originally written. As far as the CPU is concerned, the instructions are coming to it in its own language.

This approach has some drawbacks: You can't change anything in the program in the same way you couldn't change the taped version of your Italian speech. When you buy a program that's already been translated into machine language, it's called a **compiled** version of the program. The operation we've been calling translating is actually called **compiling**.[13] Another way to refer to this compiled (already translated) program is to call it the **object code** version.

Almost all the software companies supply their programs in object code form in order to keep the programs secret. They really don't want you to add a few lines of instructions and create a new version of their programs. Since it's almost impossible to decipher instructions written in machine language, the software manufacturer's trade secrets are much safer.

Programs Built into the Computer

Now let's look at a software company's ultimate in protection. Suppose the company took it's accounting program (in machine language), loaded it into RAM (memory), and then zapped the RAM, "freezing" the silicon atoms in place. This means the Annies that are standing up will stand up forever. The Annies that are sitting down will remain sitting forever. It doesn't matter if the computer is turned on or off; the silicon atoms will now retain their positions forever and can never return to their at-rest state. This has the effect of permanently storing the program in memory. This is a very different situation from the silicon atoms in RAM (memory) in which the atoms haven't been "zapped," can still move about, and always sit down when the electric current goes off.

If the concept of frozen silicon seems magical, think about

elements that you're used to—hydrogen and oxygen in the form of water. Water is a liquid and its molecules can move about. But when you cool water down, it freezes into ice. The process of freezing water is really no different from the process of "freezing" silicon, except that the change in silicon is permanent.[14]

This special "frozen" silicon now permanently holds the accounting program. But there is a drawback to this storage method. Since the memory mailboxes are already filled with immoveable atoms, these mailboxes can never be reused —no other program can be loaded into this memory. The good thing about this storage system, however, is that no disk drive is needed. Every time the computer is turned on, the accounting program will already be in memory, ready to use, without first having to load it in from a floppy disk.

Of course, computer professionals don't call these zapped memory chips "frozen Annies." They call them **read only memory** or **ROM** (rhymes with Tom). They're called this because the memory (mailboxes) is permanently filled so there is no way to write anything into them. However, since they are filled with instructions for the CPU to read and follow, the memory is for reading only, and hence its name.

Why would a software company do this? One reason was already mentioned—so the computer could be used without a disk drive. Remember, with a personal computer you need to load the program you are going to use into memory (RAM) from your floppy disk. But disk drives are expensive. A ROM chip provides a way for the CPU to run a program without having first to load it into memory from a disk.

For example, some electronic typewriters provide limited word processing capabilities. This type of electronic typewriter is really a computer—it has a CPU and some RAM. These machines may have only 100 or 200 memory mailboxes (RAM), enough to hold one or two lines of text, or as many as 3,000 or 6,000 memory mailboxes, enough to hold one or two pages of text. They each have a ROM chip that holds a simple word processing program. A typist can correct typing mistakes, insert and delete characters on a few lines of text, and when she's satisfied with it, print it out. Since this electronic typewriter doesn't have disk drives, there's no way to store any of the typed material, such as form letters and reports. When the electronic typewriter gets turned off, the text in RAM vanishes (as all the Annie atoms sit down). But the instructions that make up the word processing program don't disappear, since they are permanently stored in ROM. These instructions are available as soon as the typewriter is turned back on.

Many children use home computers in order to learn how to program in BASIC. The home computers that they use are relatively inexpensive, costing under $300. One reason they are so cheap is because their cost doesn't include disk drives (although they're often available at an additional charge). Since programming requires the use of a language program, many of these home computers include a ROM chip that contains the BASIC language program. This eliminates the expense of having disk drives (which are normally used to store BASIC until it's loaded into RAM) while still allowing home computer users to program in BASIC. These home computers use some RAM (empty memory mailboxes) to hold the program instructions the child is actually writing. However, when the computer gets turned off, the program that was written disappears from RAM. (Without a disk drive there is no place to store the newly written program so that it can be used again when the computer is turned back on.) Of course, the BASIC language program remains because it's frozen permanently in ROM.

More and more computers are being designed with application programs "built in" via ROM chips. For example, the Commodore Plus/4 home computer is designed with a circuit board that contains four ROM chips. One ROM chip contains a word processing program, another ROM chip contains a spreadsheet program, another chip contains an electronic filing program, and the fourth chip contains the BASIC language to allow users to write their own programs. This strategy is also followed by the **lap computers**, which fit on your lap (or in your briefcase). These computers are truly portable because they don't have the weight of disk drives to drag them down. Instead, brand-name business applications such as Lotus 1-2-3 or WordStar are packaged inside in the form of ROM chips.

Specialized ROM chips are beginning to show up in all types of consumer products. For example, the new cars that remind you that its time to get the oil changed have a CPU that is following a program on a ROM chip. The program simply has the CPU compare the mileage reading on the odometer to preset suggested oil change mileages.

Game Cartridges

If a program is recorded on a tape or a disk, it can be copied. If a program is on a tape, you can duplicate it in the same way you would copy any cassette tape, by using two tape recorders. A program that comes on disk can be copied to another disk. If you have two disk drives, you just use a simple copy program that reads the program from one disk and writes it to the other. However, when you use chips that have a program "frozen" into them, it is impossible to make copies of the programs.

Although the programs you buy have copyright protection and it is illegal to make copies, many people do it. Since computer companies don't want to have to sue third graders for making illegal copies of game programs, many game manufacturers choose to distribute their programs in the form of cartridges. Inside the plastic cartridge is a ROM chip with the instructions that make up the game imbedded in its memory (mailboxes). To use the game, the chip (game cartridge) is inserted into a cartridge slot on the home computer, which is specially designed for this purpose (similar to the slots in eight-track tape machines). Not only does this protect the game program from being copied, but, again, cartridges with ROM chips allow home computers without disk drives to run programs or play games that would otherwise have to be loaded into memory from a disk.

Why Aren't ROM Chips Used More?

Most business application programs are not distributed in the form of ROM chips. These programs are still distributed on floppy disks. There are several practical reasons for this. First, most personal computers don't have cartridge slots, probably because they have disk drives instead. This is beginning to change, however. The Commodore 64 and IBM PCjr come with "outlets" in which to plug ROM chip cartridges, and some of the portable lap computers are also equipped with chip outlets.

One of the other drawbacks of putting application software on ROM chips is that the program can never be inexpensively revised, as it can when it's distributed on floppy disks. Software companies continually revise their programs in order to fix "bugs" (program errors) or stay ahead of their competition. When you buy a program such as word processing, spreadsheets, or database management, you get the latest version of the program. The program you get is identified by its brand name and its **version number**. The numbering scheme is a number, a decimal point, and another number. For example, I'm writing using WordStar 3.0 (pronounced three point oh). But some time has elapsed since I purchased WordStar. Since then some wonderful new features have been added to the program. I could get the benefits of the new features by upgrading to the latest version (version 3.3). The way to upgrade is simply to pay a nominal fee to the software company and to ship them back your original disk. In turn they send you a new disk with the latest version of the program on it.

This is the most cost-effective way to be able to keep getting the latest improvements in your software. The disk itself costs the software company about $1 (when the disks are purchased in quantity) and the process of duplicating a disk is cheap and fast. In contrast, the cost of manufacturing a ROM chip is much higher, since it includes both the cost of the memory chip itself and the process of freezing the program into the silicon atoms in the chip's memory mailboxes.

Even after you spend the money to obtain a new ROM chip containing the latest version of your program, you're not done. The old ROM chip has to be removed from the circuit board and the new chip put in. Since some chips are soldered into their sockets, this may require taking your computer into a computer store and having a technician do this for you—for a fee, of course.

Software Compatibility Depends Upon the Brand of CPU

Just because you have a personal computer doesn't mean that all the programs in the marketplace are going to run on it; nor will you be able to stick any game cartridge into the cartridge slot on your home computer and expect it to work. The question of whether or not a particular brand of software will run on a given brand of hardware—that is, compatibility—depends on whether or not the program you want to use was compiled for your computer's brand of CPU.

Remember that there are different brands of CPUs and each has its own machine language. The application programs you buy, such as word processing or spreadsheets, generally come already compiled in a particular machine language (object code). If your CPU's machine language is different from the software company's, its program will be sending machine language instructions that your CPU can't possibly understand. To put this in terms of our Italian-speaking CPU, imagine taking the tape that contains the speech that has been translated into Italian and playing it to a French audience.

Compatibility Also Depends Upon the Operating System

Unfortunately, software compatibility doesn't stop with just making sure that your CPU and the CPU the software was compiled on are the same. Another factor is involved—the method for interacting with the peripherals.

Software programs contain instructions for the peripherals, as well as for the CPU. For example, the word processing program has to instruct the disk drive to go to parts of the floppy disk and either read or write information there, as well as instruct the screen to display the text and the printer to print out documents. Getting those peripherals to work with the CPU requires many detailed instructions, which is also known as **handshaking**. For example, look at the steps necessary for printing. Before beginning to send characters to the printer, a program must verify that the printer is turned on and ready to receive information. In order to do that, some dialogue must occur. A signal must be sent to the printer that means, "I'm going to send you characters. Are you ready?" If the printer is ready, it responds with a signal that means, "Yes, I'm ready." Time must be taken into consideration. If the "Yes, I'm ready" signal doesn't get sent within a certain interval of time, an error message must be sent to the screen to tell the user that she forgot to turn on the printer. A similar process must occur when each group of characters is sent to the printer in order to verify that the paper isn't jammed, the ribbon isn't used up, or the buffer isn't about to overflow. And this procedure is just for the printer. A whole other set of instructions, that is, a different **protocol**, is required for other peripherals such as the disk drive and the screen. Still other instructions are necessary to make sure there isn't a traffic jam when two peripherals try to communicate with the CPU simultaneously. Imagine if you tried to type on the keyboard while the disk drive was sending characters to the CPU, or two reservation agents tried to reserve flights at exactly the same moment. The **operating system** acts as the traffic controller that regulates the flow of the various electronic signals going to and from the CPU and its peripherals.

Just as human-like computer languages were developed to simplify programmers' ability to write instructions the CPU could understand, so too a way was found that made it easy for programmers to avoid the work of writing the detailed handshaking instructions necessary for communication between the CPU and peripherals. The answer is a master "traffic control" program called the operating system.

The operating system is composed of an assortment of **systems programs**, each of which has the machine language instructions for making the peripherals go through their various functions.[15] Just as the many machine language instructions in our multiplication program got activated by the instruction "MULT," the operating system's programs automatically get activated every time an application program requires the performance of a peripheral. All the application programmer has to do is write into her program a single command for the peripherals to act. For example, the command in the BASIC language to get something printed out is LPRINT. This simple instruction gets converted by the operating system into a huge list of instructions that verify that the printer is turned on, that it hasn't run out of paper or ribbon, that the print buffer isn't about to overflow, and so on.

Because of the operating system, the application programmer can concentrate solely on figuring out the series of logical instructions necessary to accomplish a specific task such as cost accounting. She doesn't have to also worry about the details of communicating with the hardware, because those details have been taken care of by a specialist— the **systems programmer**.

Just as there are different brands of CPUs, there are different brands of operating systems. For a program to work on your computer, your computer must have the same brand of operating system that the software company built in to their program. Remember that after a program is written in a computer language, it gets compiled (translated) into machine language. That process also integrates the operating system's characteristics into the resulting machine language program.

The Operating System Is the Computer's Nervous System

When you sit down at a computer and turn it on, don't expect that it will be ready to go to work for you yet. In fact, no matter what keys you strike, they won't be displayed on the screen. Nothing will work and nothing will happen until you load the operating system into memory.[16]

I think of this phenomenon as similar to what happens when a person has brain damage that affects her central nervous system. She may look fine, but if her doctor pricks her legs she won't feel a thing. Her problem is the neuron paths that connect her feet with her brain aren't transmitting "messages." In a similar way, a computer has a central nervous system—the operating system—and if it's damaged, or simply missing from memory, the peripherals won't be able to communicate with the CPU (the computer's brain).

Since the operating system acts as the intermediary between the CPU and the peripherals, every computer—whether it's a $99 Atari or a $1,000,000 IBM—has to have an operating system in memory (RAM) before it can be used. All the hardware in the world is of absolutely no use unless you have at least one piece of software—the operating system.

For that reason, when you buy a computer, you always get the operating system included in the purchase price. Or I should say *almost* always. Some unscrupulous retailers advertise a particular computer system at a low price and then mention in small print that the operating system in not included.

Booting: Loading the Operating System

In order to use a computer, you must first place the disk containing the operating system into the disk drive.[17] The computer's on switch is wired so that the first thing that happens when you turn the power on is that the read/write head of the disk drive moves to the location on the disk where the operating system is.[18] It then reads (plays) the operating system into memory from the disk (Figure 6-3). This takes a few seconds; until it's done the computer won't respond to anything you're typing at the keyboard and the screen will remain blank.

This process of loading the operating system into memory is called **booting the system**. The term comes from the concept of the self-made person who pulled herself up by her bootstraps. As applied to computers, the term, often shortened to **booting**, serves to remind you that all that expensive hardware is just junk until it's pulled up by the operating system's bootstrap.

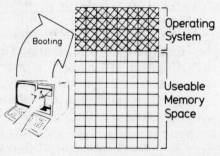

FIGURE 6-3. The operating system has to be loaded into memory before the computer can be used. To do this, the disk with the operating system is inserted into the disk drive, and the operating system program is automatically played into memory.

System Prompts

Once the operating system is in memory, some type of message will come on the screen to let you know that the computer is now ready for you to use. Unlike error messages that tell you that something has gone wrong, the purpose of this message is to prompt you into action, which is why it's referred to as the **prompt** or **system prompt**. For example, the Radio Shack operating system puts the message "SYSTEM READY" on the screen to let you know that the operating system has been booted (loaded) into memory, and the more inscrutable prompt, A>, is used by the operating systems that come with IBM and KayPro computers.

The Operating System Is Also Your Hostess

In addition to being a traffic manager, the operating system also acts like a hostess in a restaurant, seating new customers. Instead of customers, it "seats" programs, calling them into memory from the disk. When the system prompt comes up it says, in essence, "I'm ready for work. What should I do?" It is prompting you to reply with the name of the application program you want to use. If you were going to do word processing, you might type in WORDSTAR. The operating system then knows to go to the disk drive, look on the floppy disk for the WordStar program, and then load (read) it into memory (RAM) so that you can actually use it.

Everything works together. The application program communicates with the operating system, and the operating system "talks" to the hardware. As Figure 6-4 shows, this means that both the operating system program and the application program must be in memory at the same time before the application program can work.

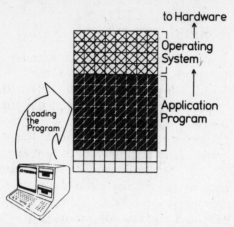

FIGURE 6-4. An application program does not talk directly to the hardware. Instead, the program talks to the operating system and the operating system talks to the hardware.

Shop for Software First

Different brands of computers use different operating systems. What does this mean to you? It means that when you shop for a program you need to notice what operating system it requires. If it was turned into machine language on an Apple computer that had Apple DOS (Apple *D*isk *O*perating *S*ystem) in memory, it will only work on an Apple II computer. If it was turned into machine language on a Radio Shack computer with TRS-DOS (*T*andy *R*adio *S*hack *D*isk *O*perating *S*ystem) in memory, it will only work on Radio Shack TRS-80 computers.

This is why you need to shop for software first. Once you find a program that's appropriate to your specific needs, for example, sales representative accounting with a particular commission structure, you have narrowed down the choice of computers that you can use. The only computers that will be able to run the program you've selected are those that use the operating system specified by that program. This sounds like once you identify the software you want, you won't have any choice about which computer to buy. Actually, that's not completely true. To see why, it's necessary to look at the way computers are manufactured.

A Computer Is More Than the Sum of Its Parts

Most people assume that if they buy an IBM computer they are getting IBM parts, and that a Kaypro computer comes equipped with Kaypro-made components. In fact, none of the computer manufacturers actually make its own parts. Instead, they buy standard parts from other manufacturers. For example, there are a few chip manufacturing companies, such as Intel, Zilog, Motorola, and National Semiconductor, that design and make CPUs. Some of the most commonly used disk drives are manufactured by companies such as Tandon, Shugart, Micropolis, and Seagate. There are also companies that make RAM chips such as Advanced Micro Devices, MicroMemory, and Plessy. Still other companies manufacture screens and keyboards.

A computer manufacturer is therefore really only a packager. Its big task is to wire together all the components that it purchased so that the CPU will be able to communicate with its peripherals. Not very long ago if you wanted to have your own microcomputer, you bought your components separately and put them together yourself. Apple was one of the very first microcomputer companies to actually do the packaging for you, which meant that suddenly you didn't have to be an engineer to own a microcomputer.

The good thing about all these computer manufacturers not making their own parts is that there's really nothing to worry about if a computer company goes bankrupt like Osborne and Victor did. Owners of those brands of computers are still able to get all the spare parts they need, because the big companies that made the parts are still in business. That's why I have to chuckle when I hear people say that they're willing to pay more for a "big name" computer because they'll always be able to get parts or because it's a name that can be relied on to make durable components. Generally, the only parts that the manufacturer makes itself are the plastic case and the wiring (circuit board).

The computer manufactuer's job of interconnecting the components may sound fairly easy, but it's actually a complex undertaking. The way that the interconnections between the CPU and the peripherals are made are based on the computer manufacturer's unique design. Even if two microcomputer manufacturers bought all the same parts, the resulting computers would be different because of the way the components were connected together. It's like two children building a toy house using the same erector set. All the parts will get used, but each child's house will be unique.

Brand-Name Operating Systems

Each manufacturer supplies you with an operating system to get its CPU and peripherals to work together. Since each manufacturer has a unique method for interconnecting peripherals and its CPU, each machine requires its own operating system. This means that even if two brands of computer use the same brand of CPU, they will have their own particular (brand name) operating systems.

The history of computers is that every company that manufactured large (core memory) computers also had a staff of programmers to write operating systems to go with them. At first things were the same with personal computers. Apple computers came with Apple DOS, Radio Shack computers came with TRS-DOS, and so on. These operating systems, which are written by computer manufacturers, are called **proprietary operating systems**.

There was a big drawback to this approach, however.

Keep in mind that applications programs will only work with one specific operating system. This meant that programs would only work on one particular brand of computer. Programs that expected TRS-DOS to be in memory would only work if TRS-DOS were really in memory. Since the copyright to TRS-DOS was owned by Tandy, this meant that some programs would only work on Tandy computers. Similarly, some programs would only work on Northstar computers, others would only work on Apple computers, and so on.

This state of affairs was no accident. The computer companies loved it. Once they sold you a computer, they had you as a captive audience, purchasing software from them as well. And they certainly wanted you to buy *their* programs, since their profit margin on software is a lot higher than on hardware. After all, to make another computer you have real manufacturing costs such as raw materials and labor. But to make another copy of a program, all you have to do is ask your operating system to read the program into memory and write it out onto a blank disk. The cost to them is about $1 for the floppy disk. Once a company has made back the money it spent on programmers to develop the software, the rest of the income is almost pure profit.

This situation also limited businesses from switching over to a different manufacturer's computer when they outgrew their original systems. When a business had a lot of money and time invested in programs that only ran on one brand of computer, switching to a computer that used a different operating system meant a huge expense in **converting** all the old programs, that is, rewriting them to work with a new operating system. Therefore, a business would almost be trapped into buying any new computers from the same manufacturer. It's almost as if General Motors could ensure that if you bought one of their cars you'd be trapped into General Motors cars forever—that when you needed a new car it would be impossible to buy a Toyota!

Programmers who wrote application programs hated this state of affairs. After all, if you'd written a great word processing program, you'd want to sell it to everyone who had a computer, not just to people who had the same computer that you did. But proprietary operating systems resulted in exactly that situation. For example, if you look at the description for AppleWriter (a word processing program), you'll see that it requires the Apple-DOS operating system. If you look at the write-up for Scripsit (another word processing program), you'll see that it needs the TRS-DOS operating system found only in Radio Shack computers. Programmers who wanted to sell their programs to people who had Radio Shack computers as well as to people who had Apple computers had to buy both computers and compile two different versions of the program. That was quite a bother, but it wasn't all that much work considering there were only a few major brands of computers in the early days of microcomputing.

Imagine the problems with trying to deal with the more than 600 brands of microcomputers that exist today. If you were a programmer writing a word processing program, you'd have to buy 600 different brands of computers in order to be able to sell the program you wrote to anyone who wanted it. That would be much too expensive; therefore, many programs were initially only available for one particular computer.

CP/M—An Almost Universal Operating System

Then everything was changed by one company, Digital Research. It had the idea of writing a tailorable operating system that would run on many different brands of computers. Its program is called CP/M, which stands for Control Program for Microcomputers (but of course no one ever says the whole name, they just use the initals—See-Pee-Em).

Think of CP/M as a sort of "adaptor"—something like the converter you can use to get your hair dryer to work in Europe (which has a different type of electricity than the United States). Unlike the electrical converter that is hardware, CP/M is software (a program that you load into memory).

It's important to remember that there are just a few manufacturers of CPUs. In fact, only five or six CPU manufacturers supply all the CPU chips for the 600 different brands of personal computers on the market today. Because most computer manufacturers use the same brand of CPU, Digital Research wrote CP/M for the most popular CPU chip at the time—the 8080 chip manufactured by Intel. (Actually the whole name for CP/M is CP/M-80—the 80 reminds you of the type of CPU chip that CP/M works with.)

Digital Research wrote CP/M in a tailorable form, which meant that as long as the CPU was the Intel 8080 chip, CP/M could be taught to work with most peripherals. It could also be installed on souped-up versions of the 8080 chip, such as the Z80 and Z80A CPUs made by Zilog. As a result, hundreds of brands of computers could use CP/M as their operating system—Radio Shack computers, Northstar, Kaypro, Osborne, Epson, and Compupro, just to name a few.

This made it possible for a programmer writing a program on a Radio Shack computer to load the CP/M operating system into memory, then load the CP/M-compatible language program into memory, and write an application program that could be used on any computer for which CP/M was available. The resulting program would be **CP/M compatible**. For example, the word processing program WordStar is CP/M compatible, so it will work on every brand of computer that can use CP/M as an operating system. As Figure 6-5 shows, the same program works on each of the three systems. The black program talks to CP/M, and CP/M "converts" the program's operating system commands into instructions compatible with each of the three computer systems—white for the white computer, dotted for the dotted computer, and lined for the lined computer.

Personal computer owners who wanted to use a CP/M compatible program merely had to go out and buy themselves a copy of CP/M and install it on their system (if their

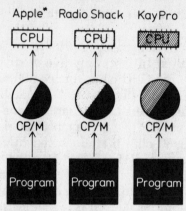

FIGURE 6-5. The CP/M operating system acts as a "converter." Here the same "black" version of the program works on different machines because CP/M has converted the machine-dependent instructions appropriately for each of these computers. Once you have the version of CP/M for your machine, you have "unlocked" your computer and are able to use the larger "library" of CP/M-compatible software. (The Apple II requires an additional CPU to use CP/M.)

computer used an 8080, Z80, or Z80A CPU chip). This ability to use CP/M meant that their software options were greatly expanded. They were no longer locked into software for one particular brand of computer; instead they could choose from a much wider range of available programs. So the jig was up. Computer manufacturers no longer had a captive audience. If you had a Radio Shack computer, you could now use CP/M for your computer and buy programs that had been written for lots of other machines.

When new companies decided to go into the computer manufacturing business, they had to ask themselves, "Why should we pay to develop out own specific operating system when our users are going to buy and use CP/M anyway?" If they supplied their own operating systems they not only had to pay programmers a great deal of money to write them, but they could no longer lock in a captive audience for their software. Unlike the older computer manufacturers such as Apple, they had no established customer base whose programs were already wedded to their proprietary operating system. And by this time, almost all software companies were writing their new programs for CP/M, making it unlikely that any great new software would be specially developed just for their new proprietary operating system. So the newer computer companies stopped supplying a proprietary operating system with their computers.

Instead, the new computer companies contracted with Digital Research for a version of CP/M that was tailored for their particular peripherals.[19] Then these companies provided it to the customer on a disk with their company name on it. When you buy a KayPro or an Epson computer, for example, they are sold with CP/M as the computer's operating system.

CP/M Isn't Completely Universal

Remember that CP/M is only available for computers that use a similar CPU chip—the Intel 8080 or the Zilog Z80, Z80A, or 8085. It isn't possible to buy CP/M for computers that use other brands of CPU chips, such as the chips that Apple and IBM use. And even if two different brands of computers use the same type of CPU chip, the exact same CP/M operating system disk that is used for one brand cannot be used on another. You have to buy a specific version of CP/M for each brand of computer because each computer company interconnects its peripherals to its CPU in a different manner. (In Figure 6-5, the dotted (Radio Shack) version of CP/M will only work with the dotted computer; it won't work with the lined (KayPro) computer.) Unlike applications software where the you tailor the system yourself when you install it, the operating system is purchased after it has been tailored to run on a particular brand of computer. For example, if you want to use CP/M-compatible software on a Radio Shack TRS-80 computer, you have to buy the TRS-80 version of CP/M. Similarly, if you want to use CP/M-compatible software on a Northstar computer, you have to buy the version of CP/M tailored specially for the Northstar computer.

Digital Research never wrote a version of CP/M for the 6502 chip—the CPU that Apple II computers use. If you want to use CP/M programs on your Apple II, you not only have to buy the CP/M program, you also have to buy another CPU. The second CPU comes on a printed circuit board, which is ready to slip into the card cage of the Apple.

What Makes Computers Fast/Faster/Fastest?

If this were 1983, we'd be finished with software and hardware compatibility. But advances in personal computers are continuing. First, we'll look at what's happened recently and where the IBM Personal Computer (IBM PC) and the Apple Macintosh fit into the picture. Then, we'll look at where things are heading and what the new trends seem to be.

In order to understand the newer, faster computers —1983 models and on—we have to go back to the design of the CPU. Keep in mind that all the instructions of a program actually get performed in the CPU. When a program requires that two numbers be added together or one charcter be compared to another, the characters first have to be moved to the CPU.

Let's suppose we want to compare a character (a byte) that's in RAM (memory) and see if it is a 4 (remember our multiply program had to do that). To make the comparison, first, we have to move the byte we want to compare to the CPU; then we have to compare it to a 4.

In Chapter 1, we used a telegraph approach, implying that the electrical impulses (bits) that represent a character (byte) travel one after another along a kind of telegraph wire to get from RAM to the CPU. But that would be quite time consuming. To see how computers achieve faster speeds, here are some alternate designs. For comparison purposes their speeds are given in imaginary clock ticks (something like a metronome).

Let's begin with the telegraph approach. If we did have a telegraph line between the CPU and RAM, we'd have to tap out each bit of the 8-bit Morse code that represented a single character. Sending a character from RAM to the CPU would require that each bit be sent down the wire one at a time, which means it would take eight ticks of our clock to send one character. In computerese this method of sending bits in sequence is called **serial** transmission. Once the 8-bit character got into the CPU (into its register), it could be compared in one tick of the clock.

One way to speed things up is to use eight wires between the CPU and memory. All 8 bits that compose a character are then sent simultaneously. In computerese this method is called **parallel** transmission. Using a parallel approach, it only takes one clock tick to transfer a character from memory to the CPU. Then the character can be compared in another clock tick. With this approach a character can be sent from memory to the CPU and compared or added in a total of two clock ticks. Obviously, this is a much faster design than the approach using the telegraph line, which took a total of nine clock ticks to accomplish the same thing.

In fact, all microcomputers designed up until 1983 used this eight-line system. The CPU's register held 1 byte (8 bits), which meant that 1 byte (8 bits) could be compared or added at one time. For this reason they were referred to as 8-bit CPUs. In computerese the clump of wires used to transport characters is called a **bus**; if 8 bits can be sent simultaneously, it's called an 8-bit bus. To conceptualize this, think of a school bus with eight seats. Depending upon which character we're transporting, one or more seats will be occupied. For example, to represent an "A" (which we've been representing with a binary code of 01000001), two specific seats would be taken—the second and eighth seats (see Figure 6-6).

Both the bus and CPU designs have been improved, and there are now several different types of computers available other than the original 8-bit standard.

16-Bit Computers

Imagine a computer that's designed so 2 bytes can be compared at the same time. Its CPU's registers will be "wider"— wide enough for 2 bytes. Let's give this computer a wider bus too—one with 16 seats. Naturally, this computer will have 16 wires in the bus connecting the CPU to RAM. In this design, two characters can be moved in one clock tick. The computer we are describing *used to be* the definition of a **minicomputer**, that is, a computer with a 16-bit CPU and a 16-bit bus. Microcomputer technology has evolved, however, and there are now microcomputers with these same

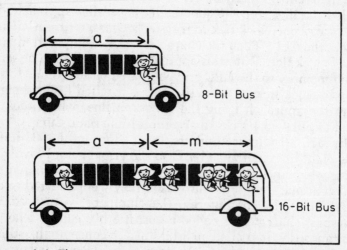

FIGURE 6-6. The representations of characters move between the CPU and memory on a bus. If the bus has 8 "seats," then one byte can move at a time. If the bus has 16 "seats," 2 bytes can move at a time.

16-bit attributes, which has blurred the old definitions between minicomputers and microcomputers.

To understand the practical differences between 8- and 16-bit machines, let's give our two computers a task to do. Suppose that we're doing word processing and we want to find every occurrence of the word "american" and change it to "American" (see Figure 6-7). To do this each byte of text

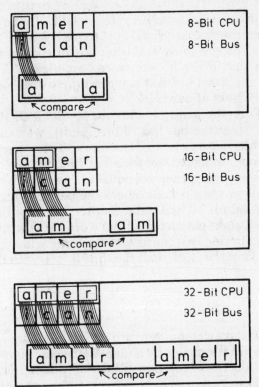

FIGURE 6-7. The architecture of a computer determines how fast a given job will take to complete. Here, a word processing program is performing a search-and-replace in order to replace every occurrence of american with American. In one cycle, the bus transports character(s) from memory. In another cycle, the CPU compares the character(s).

must be transported from RAM to the CPU's register where it will be compared to determine if it is forming the word "american." Let's time how long this particular process will take using each of the computers.

First, let's use our 8-bit CPU, 8-bit bus computer. In the first tick of the clock the 8 bits forming the first character (byte) are moved down the bus in one burst from RAM to the CPU. It takes another tick of the clock to compare that byte to the letter "a." If it is an "a," we will have to move the next byte in the word from RAM to the CPU. In the next tick we could compare that second byte to see if it is an "m." We will have to do this move and compare for each of the eight characters that make up the word "american." To determine that the word in RAM is indeed the word "american" will take 16 ticks of our clock.

Now let's do the same job with our 16-bit CPU, 16-bit bus computer. Since the bus has 16 bits (seats), we can move 2 bytes (16 bits) in one tick of the clock. Then, because the CPU has room in its register for all 16 bits, we can compare the 2 bytes and see if they are equal to "am." This will take another tick of the clock. To check the next 2 bytes to see if they are equal to "er" will take another 2 ticks of the clock. To check all eight characters of the word "american," moving and comparing two characters at a time, will take a total of eight ticks of the clock, half the time it took with the other computer.

32-Bit Computers

We can also have a computer with a 32-bit CPU and a 32-bit bus. That's what's meant by a **mainframe** computer such as the IBM 370. Those computers take only four ticks of the clock to do our task, since they check 4 bytes at a time.

Not very long ago you could distinguish among microcomputers (personal computers), minicomputers, and mainframe computers by the size of the CPU's registers and the width of the bus—8-bit CPU, 8-bit buses were microcomputers; 16-bit CPU, 16-bit buses were minicomputers; and

32-bit CPU, 32-bit buses were mainframe computers. But personal computers now come in a variety of designs with larger and larger CPUs and with different bus sizes. Let's look at some popular brands to put things in perspective.

IBM PC—Intermediate Design

The IBM PC uses an intermediate strategy. It has a 16-bit CPU but only an 8-bit bus. In doing a compare of the word "american," the bus goes to memory, gets the first byte, moves it to the CPU (one clock tick), goes back to memory, and gets the second byte (another clock tick). When the 2 bytes are in the CPU, it can compare them to the characters "am" in one clock tick. It takes a total of three clock ticks to check the first two characters. To check all eight characters, it takes 12 clock ticks, which is faster than the oldest micro-computers but slower than 16-bit bus computers.

When IBM announced the architecture of its PC, most computer specialists assumed that it would never sell. After all, other microcomputer manufacturers such as Tandy were selling faster 32-bit CPU computers for about the same price. But those computer professionals forgot that most people who buy computers never heard of buses or even CPUs. For the most part they knew nothing about comput-ers and were relieved to be able to buy a "big name." (Imag-ine how you would feel if you didn't know very much about computers but were given the responsibility of choosing the brand of computer for your company. You might be buying a few hundred personal computers for all the middle man-agers in the company, and an error in judgment would be disastrous. If you bought IBM computers, you'd be safe. If anything went wrong, you could say, "I bought the best. Whoever expected anything to go wrong with IBM ma-chines?")

Tandy 2000

The Tandy 2000 uses a 16-bit CPU and a 16-bit bus. If you watch a program run on the Tandy 2000 and then watch the same program run on an IBM PC, you'll be struck by how much faster it seems to go on the Tandy 2000. Using our compare example as a test—(computer professionals call such a test a **benchmark**)—the Tandy machine would only take 8 ticks of the clock, as opposed to the 12 ticks of the clock for the IBM PC. The Tandy machine is 50 percent faster.

Apple Macintosh—The Fastest

At the same time that IBM was coming out with its PC, Apple was developing its Macintosh computer. The Macintosh uses a Western Digital 68000 chip for the CPU, which is a 32-bit CPU with a 16-bit bus. Again, this is an in-between design. If we were going to do our search for the word "american," it would take less time than the 16-bit CPU system but more time than a computer that had a 32-bit bus as well. The Macintosh would get two bytes, move them to the CPU (one clock tick), get another 2 bytes (one more clock tick), then compare the 4 bytes to "amer" (one more clock tick). In the same way, it would compare the next 4 bytes. To compare all 8 bytes would take six clock ticks—twice as fast as the IBM PC.

As of now, there's no microcomputer version of the 32-bit CPU, 32-bit bus. That architecture is only available on mainframe computers. So for now the old definition of mainframe remains true, although it's doubtful that it will be true for very much longer given the breakneck speed of technological advances in this area.

MS-DOS—A Newer Standard for Operating Systems

Looking at Table 1 you can see various computer architectures represented in each column. These columns also break computers into "families." For example, the 8-bit CPU, 8-bit bus computers are CP/M computers. The CP/M operating system was designed for this particular architecture——there's no way to run CP/M on computers with any other designs.[20]

When IBM chose the 16-bit CPU, 8-bit bus design, it decided not to write its own proprietary operating system. Instead, IBM hired a software company, MicroSoft Corporation, to write an operating system for its computer. Using a strategy for these new 16-bit CPUs similar to the approach Digital Research had used for 8-bit CPU chips, MicroSoft wrote a tailorable operating system and retained the copyright to its program. MicroSoft named their operating system with their initials—MS-DOS (MicroSoft Disk Operating System). When you buy an IBM PC computer, you get an operating system with a label on it reading PC-DOS. But it is actually just MS-DOS tailored to the IBM peripherals. This is like buying a Kenmore washing machine from Sears—it's actually a Whirlpool washer that Sears put its private label on.)

What is UNIX?

Just as CP/M is the standard operating system for 8-bit computers and MS-DOS is the standard for 16-bit computers, there will be a standard operating system for 32-bit computers. Many people think that the UNIX operating system will become that standard.

Table 1. Computer Families

Architecture	8-bit CPU / 8-bit Bus	16-bit CPU / 8-bit Bus	16-bit CPU / 16-bit Bus	32-bit CPU / 16-bit Bus	32-bit CPU / 32-bit Bus
Definition*	Microcomputer		Minicomputer*		Mainframe*
Timing†	16 cycles	12 cycles	8 cycles	6 cycles	4 cycles
Brand Names	KayPro 2 Epson QX-10 Morrow Radio Shack Model 4, 12 Apple IIe, IIc Osborne 1	IBM PC Compaq Portable Televideo PC Sanyo Columbia Zenith Z150 Tandy 1200 Data General 1	AT&T PC Compaq DeskPro Tandy 2000	Apple Macintosh Radio Shack Model 6000 Onyx Fortune Wicat	Still in development
Major Operating System	CP/M‡	MS-DOS (or PC-DOS)	MS-DOS	UNIX (or Xenix)	
CPU Model Number	Z80, Z80A‡ 8080	8088	8086, 80186	68000, 28000	

*These definitions are no longer effective. At one time microcomputers were computers that had a particular architecture—8-bit CPU and 8-bit bus. Continued development of silicon chip-based computers has rendered the old definitions of mini-computer and mainframe computer obsolete.

†Timing is given for the task of comparing an 8-character word to determine if it is equal to a predefined word. For example, how long will it take to determine if the 8 characters in memory are really the word "american"? These values are for comparison only.

‡CP/M will only work the Z80 family of CPU chips—including the 8080, the Z80, and the Z80A. To use CP/M on the Apple II family of computers requires the addition of a second CPU from this family.

There is a UNIX operating system developed by MicroSoft and tailored to run on the Radio Shack Model 16, called Xenix. Theoretically, you can run programs on it that have been developed on other UNIX computers. I say theoretically, because there are currently so many versions of UNIX that there really is no single standard yet. For example, AT&T manufactures a 32-bit computer that uses the UNIX operating system, but it uses a different version of UNIX called UNIX V. IBM has introduced a UNIX operating system, PC/IX, which is only compatible with UNIX III. Still another UNIX operating system, developed at the University of California at Berkeley, called (appropriately) Berkeley UNIX is competing in the race to see which version of UNIX becomes the standard.

Right now, no one knows which version of UNIX will become the one that all the software developers will choose to use. Until a clear favorite emerges, each software developer will make an independent decision about which version of UNIX to develop software for. That's the current state of affairs: There's not a whole lot of programs available for UNIX computers, and there's no common library of software that will work on all the new 32-bit microcomputers. Instead, just like the old days, there is software that works on each brand of hardware. Once again, the software is hardware dependent, and buyers of the newest machines are trapped into a limited pool of available programs. As 32-bit computers become more popular, it's likely that better standardization will occur and bring with it more options for the users of these machines.

The impact of this lack of standardization was experienced by buyers of the newest 32-bit CPU computers such as the Macintosh. For example, when the Macintosh was introduced there was just about no software available for it (except for a few programs developed by Apple). After all, software developers couldn't begin writing programs compatible with the Macintosh's proprietary operating system until they could first buy a Macintosh.[21] There were many disappointed writers who bought Macintoshes only to discover that the limitations of the Macwrite word processing program included a maximum size limit of 10 pages; what's worse is that there were no other Macintosh-compatible word processing programs to choose from! This illustrates the fact that most people who buy computers don't understand the rule: *buy your software first.* In 1984, over 250,000 Macs were sold—in spite of the fact that no real business applications were available for them yet! Of course, as the years go by, the library of Macintosh software will grow. Already a new version of Macwrite is available; one that permits much longer documents.

Some companies get around the problems of introducing new computers that have no compatible software by packaging two CPUs in their computers. For example, the 32-bit Radio Shack model 16 also has an 8-bit CPU in it. This enables model 16 owners to use the vast library of CP/M software while waiting for the new generation of 32-bit software to be developed.

Similarly, the DEC Rainbow includes two CPUs. But in this case, they are a 16-bit CPU that MS-DOS works with and an 8-bit CPU that CP/M works with. Owners of DEC Rainbows have their cake and can eat it too—they can run either MS-DOS programs or CP/M programs. It is sometimes possible to buy CPUs as add-ons. For example, you can buy an Apple DOS-compatible CPU for your IBM PC. By adding the second CPU you could run both Apple-compatible and IBM-compatible programs on a single machine.

Operating Systems for Multiuser/Multitasking Systems

The first wave of microcomputers were all **single-user** systems, which meant that the CPU could only handle one keyboard and screen. Because of this, two or more people couldn't work at the computer at the same time. More recently, advances in CPU and bus design have given personal computers so much additional capability that they can support several users simultaneously. This is accomplished by using special operating systems that are designed to permit more than one user to communicate with the computer system. The operating system acts as traffic manager, making sure there aren't collisions when two users try to do something at exactly the same moment.

You see **multiuser systems** when you make airline reservations. But if a computer has two terminals connected to it, two travel agents can both try to make a flight reservation at exactly the same moment. How can the system handle that? Without a multiuser operating system there would be no way to prevent the same sort of problems that happen when two people talk at once on their CBs or two cars enter an intersection at the same time.

Because it takes more complicated instructions to handle the problems associated with many users, multiuser operating systems are much longer than single-user operating systems. As a result they require a lot more RAM—often as much as 128K RAM for just the operating system alone. (Remember you still need room in RAM for your application program and some data.) Since multiuser systems require significantly more RAM than single-user systems, why not just let each user have her own separate computer with 64K RAM?

The primary reason for having a multiuser system is so that several users can share the same filing cabinet, that is, the same disk file. Think of the airline reservation system. As soon as you reserve an airplace seat, the file is updated to show that there is one less seat available on that flight. It would be absolutely useless for each reservation agent to make reservations on a single-user computer, because each person's data file would only reflect her own sales and wouldn't show what seats the other agents had sold. Similarly, in many office situations more than one user may want to use information contained in a single file. The sales department may want to call up the inventory file to see if an item is in stock, and the warehouse will be updating the file as goods are removed and received.

UNIX is a multiuser operating system. But you don't need a 32-bit CPU to have a multiuser system. There are some 8-bit operating systems, the most common of which is called MP/M (Multi-User CP/M). The two best-known computer manufacturers whose machines run MP/M are Compupro and Altos.

Multiuser systems are also **multitasking** systems. That is, each of the users can be running a different program while still sharing a common data file. In the office, the bookkeeper may be posting payments to clients' records, the secretary may be sending form letters to the same clients using word processing, and the manager may be using a spreadsheet program to prepare next year's budget based on this year's client payments. All have access to the client file although each user is performing different tasks. However, there is no reason for the bookkeeper, secretary, and manager to use a multiuser multitasking computer system unless, as in this example, they all wanted to use the same data file at once.

There's still one more type of operating system—the **single-user/multitasking** system. Regular single-user systems can only perform one task at a time. If you're having the computer sort all the names of your mailing list into zip code order, you cannot use it until the sorting is complete or you totally **abort** the run (obviously no woman thought up this term for stopping a program). A single-user/multitasking system means that only one person can communicate with the computer, but that person can do more than one job (task) at the same time. For example, you can put the computer to work printing form letters; then when the phone rings with a client inquiring about her bill, you do not have to tell her that the computer is busy. You can just look up the information without disrupting the printing. To you it looks like the computer is doing two things simultaneously, but actually the CPU is doing one thing at a time. In this example, the printer is so slow that while the CPU is waiting for the printer to print a line of text, it can look up information stored in the database. The most common multitasking operating system is called Concurrent CP/M (CCCP/M).

Multitasking operating systems have one big limitation——users can only see one task displayed on the screen at a time. This means, for example, that you can see the letter you're writing *or* your database record, but not both simultaneously. If you want to put some information about the person to whom you're writing into the letter, you may have to call up the database display, copy the information down using paper and pencil, go back to the letter, and continue writing, using this information.

This viewing problem was solved by Apple. The Macintosh's operating system displays each task in a **window**, which is just a small portion of the screen. When describing the Macintosh, the screen is referred to as the user's **desktop**. Each window is compared to a sheet of paper on the desk. This means that documents, spreadsheets, and database records can all be displayed simultaneously, even overlapping each other if the desktop gets crowded.

Success breeds imitation. New products extend the capabilities of MS-DOS to provide the same windowing capabilities that the Macintosh has. Rather than rewriting MS-DOS, the extensions that provide multitasking and windowing are offered as separate programs that work with the operating system. For example, Figure 6-8 (see next page) shows the type of display possible with the MicroSoft program, Windows. As you can see, word processing is displaying text on the left of the screen, while a spreadsheet program is active at the right of the screen. In addition, a Date program displays the calendar. Competing products such as IBM's Topview provide similar extensions to MS-DOS.

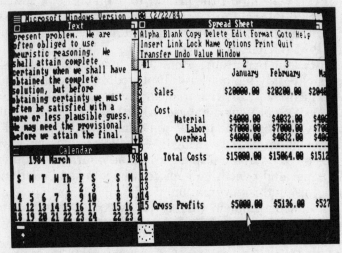

FIGURE 6-8. Windows let you simultaneously see displays from many applications. Here word processing output, a spreadsheet display, and a clock-calendar program's output are all viewed at once. (Courtesy of Microsoft Corporation.)

Look-Alike Computers

You will often hear certain computers advertised or referred to as a **look-alike** of one of the leading brands. This term simply means that the look-alike uses the same CPU and the same operating system as the brand it's imitating. By using the same operating system, the look-alike can run the same software as that developed for the well-known computer. For example, Franklin Computers manufactured an Apple II look-alike. It used a 6502 chip for the CPU and an operating system almost identical to Apple DOS. People who bought Franklins could buy and use any software that had been developed for Apple IIs. (Unfortunately for Franklin, they also copied Apple's patented design[19] for connecting the peripherals and the CPU, so Apple sued it for copyright infringement, forcing Franklin into bank-ruptcy.)

The IBM PC also spawned a host of look-alikes such as Compaq, Eagle, Columbia, and Corona computers, all of which use the MS-DOS operating system that IBM popular-ized. These companies sell their machines for substantially less than IBM, attracting customers who want to get the benefits of using software that was developed to run on IBM computers without having to pay for an IBM computer.

Some Programs Won't Work on Look-Alikes

If two computers have the same brand CPU and the same brand operating system, are the computers identical? For some brands of computer the answer is yes, they are essen-tially identical. Computers such as Compaq and Columbia will run just about every program that will run on an IBM PC. For this reason these brands are called 100 percent IBM compatible. For other brands, the answer is more quali-fied. These computers are *somewhat alike*.

Although it's generally true that two computers with the same brand of CPU and the same operating system will be able to run the same software, some software doesn't play by the operating system's rules. So although it may seem like owning an IBM PC look-alike is every bit as good as owning an IBM PC, there are certain cases when that's not quite true.

Remember, operating systems were invented so that applications programmers could write software without having to write in the detailed instructions required to com-municate with each piece of peripheral equipment. Instead, the programmer simply commands that the operating sys-tem take over at certain points to instruct and oversee the pe-ripheral's work. This makes it easy for application programmers to do their work, but there is a drawback—speed. There is an element of delay while control gets passed back and forth between the applicatons program and the operating system. Accessing the peripherals by invoking the operating system as your helper is sort of like the child's game Mother May I, where you have to ask permission before you can take a giant step instead of just going ahead on your own and taking it.

In most applications programs, operating system delay is not a factor. Graphics programs, however, are often affected by it. Because these programs are constantly sending information to activate the numerous dots (pixels) on the screen, many graphics programmers have found that for maximum results it is best to bypass the operating system and communicate directly with the screen. In essence, they write their own minioperating system for communicating with the screen as part of their application program. When a programmer does this, her program will generally only run on the exact same brand of computer that the program was designed for. These programs are called **hardware dependent.**

What this means is that if you're interested in buying a computer to run a specific program, you have to find out if the program is hardware dependent. For example, suppose at work you use the application program Lotus 1-2-3 on an IBM PC and you want to buy a computer so you can take your work home. Since Lotus 1-2-3 is a hardware-dependent program, it won't work on some of the IBM PC look-alike computers. The Lotus Development Company, the software manufacturer, has written special versions of the program for several other brands of computers but not for every one.[22]

One way to tell if a program is hardware dependent is to see if the program's description says it requires 100 percent compatibility. Some IBM look-alike computers *are* 100 percent compatible. These look-alikes have so closely copied the IBM PC that even hardware-dependent programs work on them. At the other end of the spectrum is the Sanyo. Although the Sanyo uses the same operating system as the IBM PC (MS-DOS), it's not hardware compatible with the IBM. Many programs, such as Lotus, that work on the IBM won't work on the Sanyo. Saving money by buying an IBM look-alike and still having *complete* software compatibility with the PC is possible, but it does require careful shopping.

Decisions, Decisions, Decisions

With all these choices, what should you do about buying a computer? First, you have to let go of the idea that you can wait and then get the best computer. The technology is changing so fast that no computer will be "the best" for very long. The only way to decide on whether you should get a computer now or not is to evaluate how much time and money you'll save if you had a computer doing the things that you now do by hand. Think of pocket calculators. When electronic calculators first came out, they were quite expensive. But they were still less than the old bulky adding machines. When I bought a calculator in the 1970s they cost about $35. I got good use out of it for a long time. When its batteries wore out and I tried to replace them, I discovered that the new batteries cost $10—about the same price as a new and better calculator! So I threw my old calculator away and bought a new one. But I wasn't sorry that I had gotten it when I did. It saved me many hours of adding things up by hand—it was still cost effective.

In the same way, you need to figure out the value of using a computer. Most small businesses discover that a computer pays for itself in a year or two. For example, if you maintain a mailing list at your office, figure out how long it takes to do the address corrections and the sorting by hand. Then compare it to the time it takes when you use a computer. Just that application alone may justify the expense of purchasing a computer. Estimate the percent return you get on mailing out personalized form letters to prospective clients. The result of one more sale per month may be enough to justify both the cost of the computer and the expense of doing the mailing.

I'm not suggesting that after a few years your computer

will be obsolete and you should throw it away. On the contrary, if you do a careful analysis of what your information management needs are, your "obsolete" computer will continue to give you good service even though faster, more efficient machines have been invented. For example, this book was written on a computer that uses CP/M. Many people consider CP/M obsolete since the latest software, such as Framework, won't work with it. So what? For my writing needs, CP/M is perfectly adequate. Buying the latest multiuser/multitasking UNIX computer just to write a book is like buying a Ferrari to go grocery shopping. There's no sense in paying for power that you never use!

Finally, don't look at the total purchase price of a computer and panic—look at the monthly cost. After all, if you considered your rent in terms of an annual expense you might also feel faint. When you see the cost of a computer in terms of monthly payments, it seems affordable. Typically, a small business computer system will cost under $100 a month. And once it's paid for in 3 to 5 years, you have a machine that will continue to work for some time to come—for free.

Okay, now which brand to buy? You might think that after my whole technical discussion I'd advise getting Macintosh, or Radio Shack model 16, or another 32-bit CPU system. But I don't. My answer is, "It depends." I advise getting the least expensive computer that can reliably get your work done.

Least expensive is pretty obvious. By doing price comparisons, you can figure it out. But remember, there's no way you can buy a computer for a price that will look good to you a year from now—prices just keep going down. But as my mother says, "In the long run, we'll all be dead." If you keep waiting for prices to stop dropping, you'll never have a computer. The one thing I don't advise is buying a computer from a mail order company. You'll need to develop a relationship with a local computer store so that you can get help when things go wrong, as they always do. That type of personal help isn't available from mail order companies. Whatever savings you might glean will not prove very valuable when your system won't work right and there's no one you can call for advice. You may end up paying for a service call and waiting a day for the repairwoman to get there when it's not a hardware problem at all, as the computer store would have likely told you in the first place.

The price of a computer depends somewhat on its speed. An 8-bit computer usually costs less than a 16-bit computer, and a 32-bit computer is the most expensive. But is the speed worth the increase in price? Again, it depends.

If you're buying a computer to do word processing, you won't notice much difference among computers. With word processing your typing speed is so slow that you won't really gain much benefit from a faster computer. But if you're planning on using your computer for sophisticated database management, you will notice a significant difference in speed among various models. But, again, how much will it be worth to you to increase the speed? If it takes 30 minutes to do a particular operation, it may not be worth an extra $2,500 to get a computer that will do the task in 15 minutes. But if an 8-bit computer will take 2 hours to read the 50,000 names on your subscriber list, it may well be worth the money to do the job in half the time.

How Much Memory Is Enough?

Remember, when you use your computer, the operating system and the program are both in memory. Different operating systems use a different amount of space. CP/M only takes up about 20K, but MS-DOS uses almost 64K. Multiuser operating systems may require 128K RAM or more.

When buying a computer, 64K of RAM is enough to use almost all the programs written for CP/M. Since most 8-bit computers have a maximum capacity of 64K RAM, CP/M software developers wrote programs with that limit in mind. These programmers knew that with CP/M taking up 20K, there will only be 44K left in RAM for their programs. However, many sophisticated programs have been written for CP/M that are actually larger than 44K. These programs work because the software company broke the entire program into separate segments, called **overlays**, that are no larger than 44K each.

I like to think of overlays in terms of my recipe for chicken crepes. That recipe has three parts: the crepe pancake itself, the chicken filling, and the white sauce that goes over it. Taken together, all the parts are necessary to make the recipe; however, each segment is really a separate set of operations. When I'm working on the pancake, that's all I do. I don't work on the filling or the white sauce.

Similarly, an application program like a database manager can be broken up into separate overlays—one for adding new records and others for sorting, indexing, printing a file, and so on. These segments are entirely separate, and when the applications program requires one of these operations to be performed, the overlay is brought into RAM where it does its work. When it's finished, RAM is reused, and the next overlay is brought in to RAM from the floppy disk.

Unlike 8-bit computers, the newer 16-bit computers aren't limited to 64K RAM. Because of this, many software companies now simply write their programs as one long "recipe," without breaking them into separate overlays. This means that MS-DOS programs often require 256K RAM or more to hold the application program alone. Since the MS-DOS operating system uses almost 64K itself, you may need as much as 320K or 512K RAM in order to hold the operating system, your application program, and some data.

Which Operating System Should You Buy?

Which operating system should you buy? Again, it depends. Some people refer to CP/M as obsolete because the newest software is now being developed primarily for MS-DOS. But they overlook the fact that CP/M has the biggest library of available software. So if there are CP/M programs that you know will take care of your information management needs, why pay more for a computer that runs MS-DOS? As I've said before, always shop for software first, particularly if you have highly specialized needs (such as an accounting system for thoroughbred horse breeders). There may be a CP/M program that works perfectly, whereas a similar MS-DOS program may not have been developed yet.

If you're a writer, your best bet might be a Kaypro 2 or a Morrow. These computers come **bundled** with software. This means they come with software included in their purchase price. For example, a spreadsheet, a word processing program, a database program, and the BASIC programming language are all included in the purchase price of the Kaypro. By contrast, if you buy an Apple IIe you have to pay extra for each of these programs. The software for the Apple might cost as much as $1,000!

Since the Kaypro 2 can't run Lotus 1-2-3, it would definitely not be your choice if you were a financial analyst and needed that program for your work. Obviously, there's no best computer. It depends on *why* you're buying a computer.

Shop for Limitations

Once you've identified the programs you need, you've narrowed down your choices for computers. You'll only be able to buy a computer that uses the operating system that your software requires. Within those brands of computers, you'll be limited by which peripherals will work with particular systems. Remember the example of the Sanyo? Although it's an IBM look-alike, you can't connect a hard disk drive to the Sanyo so it wouldn't be a good choice if you were planning on setting up a large database system.

The rule of thumb to follow is to determine the computer's limitations to see if they will effect the way in which you plan to use your system. Ask about what peripherals you are limited to connecting. And remember to plan for your 5-year data needs. You may not need a hard disk now, but if your business is growing or you have plans for larger record-keeping projects, you may want to make sure the system you buy today is expandable and can be upgraded in 1 or 2 years when you're ready to do so.

CHAPTER 7

Housekeeping

Working with a computer really isn't that difficult. Computer concepts are similar to things you already know: A program is a recipe, data are ingredients, and a disk is like an electronic filing cabinet that holds your data. But there are some fundamental differences in the way you work with electronic information that dictate learning new work patterns. Regardless of what applications programs you use, there are certain procedures that are common to all computers and must be followed if you want to stay in control of your data effectively.

I'm always amazed that although many software packages include a tutorial program on how to use that particular application, usually nothing is mentioned about the basic procedures that should be followed to avert potential data loss or other such catastrophes. Nor do they teach the new computer user how to manage electronic information in an orderly way. And yet, it is only by learning how to manage electronic information in a particularly "tidy" fashion that disasters can be avoided. This chapter will introduce you to the practical side of computers.

System Utilities

The basic data maintenance functions are performed by the operating system. Although the operating system primarily serves the CPU, enabling it to communicate with the peripherals, there are a few things that it does for the user as well. These include showing you the "table of contents" of a disk, showing you how large (how thick) each file is, copying files, erasing files, and renaming files. These functions are so useful they are referred to generically as **utility programs**, or sometimes **system utilities** (since they are performed by the operating *system*).

It would be hard to use your computer without the utility programs that are included as part of your operating system. They perform what is often referred to as the **housekeeping functions**, and, like vacuuming and dishwashing, these operations are done almost every day. And, like real housekeeping tasks, their importance shouldn't be underrated.

System utilities perform functions that you usually take for granted when you deal with manual recordkeeping. For example, if you want to see if you have enough empty space to fit a new file into a file drawer, you just open the drawer and look in—no big deal. But with magnetically recorded information stored in an electronic file cabinet (a disk), you can't use your eyes to look, any more than you can see what's recorded on a tape, because the data and the files can't be visualized. Instead, you have to use one of the operating system's utility programs to find out how much room is left on the disk.

Although the utility functions are easy to learn, most operating system manuals assume you already know what the utility is for. They will always describe *how* to use the utility program but rarely explain *why* you'll want or need to. To demystify these day-to-day operations, let's briefly review what these functions are for.

Keep in mind that these utilities are separate programs that are part of your operating system. Since a different team of system programmers wrote each operating system, each has its own peculiar way to accomplish these functions. For example, one operating system uses the comman ERA to erase files, another system uses the command DEL, and a third uses KILL but all the commands mean the same thing. But it's not important to go into the differences among the various operating systems in calling up these programs.

Rather, this chapter will tell you what these utility programs are all about and why you'll want to use them. This means that if you want to use someone's computer, the first thing you need to do is borrow the operating system manual and look up the particular way to perform the housekeeping functions.

Looking for the File You Want

In the same way that you can't look at an unlabeled phonograph record and know what songs are recorded on it, you can't just look at a disk and know what programs or data files are on it. Each disk should have a gummed label that is used to write down what files are stored on it. But people often forget to write the name of the files on the label, particularly if they are re-recording and the original label is filled with a list of the former files. I often forget to do this with cassette music tapes simply because a pen or pencil isn't handy. Writing on computer disk labels can only be done with a felt-tip pen, because the pressure of writing with a regular ballpoint pen will crease a floppy disk and permanently damage it. If a felt-tip pen isn't nearby, people often neglect to add new file names to the disk's label.

As a result, you are often faced with the problem of trying to locate a specific file such as a form letter you've written. If you're like me, you've looked through all the labels on the floppy disks but still can't find the file! Fear not, the **disk directory** will solve the problem. Think of the disk directory as a table of contents for the disk in the same way that a record label is a table of contents for a music album. If you want to know every song that's on a Barbara Streisand record, you look at the label. If you want to know the name of every file that's on a disk, you look at the disk directory. Actually, since the disk is really like an electronic filing cabinet, the directory is more appropriately like the white card on the

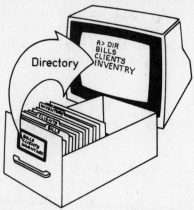

FIGURE 7-1. Think of your disk as your filing cabinet. The disk directory shows you the list of files in the cabinet in the same way that the white card on the front of a real filing cabinet does. The operating system of most personal computers restricts the length of the file name. For this reason our inventory file had to be called INVENTRY.

front of a metal filing cabinet (see Figure 7-1). At your request, the disk directory will display on the screen the names of all the files, just as the white card on a file cabinet lists all the files inside the file drawer.

Your disk might hold different kinds of files. One file might be a program such as WordStar, and another might be a data file such as a manuscript. In fact, your disk has room for many files. Depending upon the way you work, you may have one giant file containing 10,000 names and addresses, or you may have many files—lots of programs and small data files—on one disk. It's no different from a drawer in a file cabinet that can hold either one big file or a combination of small and medium-sized files.

To request the operating system to display the directory on the screen, you merely type in the command DIR. This particular command works with all the operating systems to activate the directory utility program. Once that command is received, all the names of every file that are recorded on the disk get displayed on the screen.

Maintaining the Directory

If there were any way to write a file onto a disk and somehow forget to add its name to the directory, there would be no way to ever find it again. We've all had that problem with our manual filing cabinet. We file something and forget to write it on the file cabinet white card. With a filing cabinet there's something we can do; we can open the drawer and manually read through the files until we finally locate the forgotten one. But with electronically recorded information on a disk, there's no such way to leaf through the disk looking for lost files.

Fortunately, the operating system protects us from putting files on a disk without recording them on the disk directory at the same time. As soon as a new file is written on the disk, the operating system automatically adds its name to the disk directory without you having to command it to do so.

Naming Your Files

We've talked about looking in the directory for the names of the files on a disk, but how do the files get named in the first place? Different operating systems have different methods for letting you name the files. This is similar to the differences among the various brands of manila file folders. When you buy manila file folders, you have some choices. Some folders come with wider tabs so you can put longer file names on them; others are so short that you are limited and have to abbreviate the names on the tabs.

Similarly, the various operating systems have their own rules for naming files that limit what you can call them. Both MS-DOS and CP/M have the same naming scheme.

The rule is that a file has a first name and the option of a last name. The first name cannot be longer than eight characters and must not have any spaces. The last name cannot be longer than three characters. Humans' first and last names are separated by a space; computer first and last names are separated by a period. You always provide the first name of the file, either when you begin work, when the file is created, or when you finish your work and it's time to store it on the disk, depending on the convention of the particular application program you're using. Often you can provide the last name as well; however, there are many times when the program that's being used will automatically add the last name to a file itself.

Human last names denote family connections, and people with the same last name are apt to be related. Similarly, computer last names are used to describe the type of file. For example, files with the last name BAS are programs written in BASIC. If you are programming an application for billing, using BASIC, and want to save what you've written, you must first give your program a name. Let's say you call it "BILLING." After it's been saved on the disk and you take a directory of that disk, you will see that BILLING.BAS is listed. The last name was automatically added by the BASIC program.

Another common last name is .BAK, which means that this is a backup file. Many applications programs create backup files that are simply the last version of any file that you have edited or changed. If you've worked on a file more than once, the disk will contain both the current version and the last preceding one, with the last name .BAK. On the disk on which I am working for this book there is CHAPTER4 and CHAPTER4.BAK. This means that if I do something horrendous like accidently erase a page I'm working on, I can always retreat to my backup file where the last version I'd worked on is saved.

When using word processing programs, I often assign last names myself. For example, when I write an article I give it the last name .TXT for text. When I am word processing a letter, I name the file .LTR. This way, when I call up the disk directory and see, for example, PUBLISHR.LTR I know that file is a letter to my publisher. If I didn't give it a last name, it would just appear as PUBLISHER. When referring to people, last names are often called surnames. When referring to computer files, last names are often called **extensions** or **file types.**

Checking Space Left on a Disk

Since you can't see the amount of empty space left on the disk with your eye, the operating system "tells" you this information. To find out how much space there is, you type in the name of the appropriate utility program. Although the name of this utility differs, depending on the brand of operating system you're using, all operating systems come with this function. On CP/M computers such as Kaypro, the command is STAT. On MS-DOS computers such as IBM PC, this information is included on the screen that's displayed by the DIR (display directory) utility.

It's a good thing to check the space remaining function first, when you're starting to work on your computer. If you forget to do this and the disk is almost full, you can be working away but there won't be any place to store what you've done. With no place for it on the disk, you'll end up permanently losing most of that session's work. This seems to happen to everyone at least once, after which you never forget again. Unfortunately, this is a painful way to learn.

Let's look at what actually happens when this data disaster occurs: Suppose you are writing a report. In memory (RAM) you have the operating system, the word processing program, and the text of the report you are writing. In an hour or so all the RAM gets filled by the new material you've just written, and the word processing program tries to send the text out to disk to save it. However, there's no room on the disk, and only then do you get an error message of the type "ERROR, DISK FULL."

In many systems, that disk full message is like the red light on your car that tells you that your engine has overheated. By the time the red light goes on, it's too late—you've already blown your engine. Similarly, most word processing systems have no way to recover if the disk is full—you just lose the page (or pages) in memory. Remember, most systems have enough memory space (RAM) for quite a few pages. So you may lose three, four, or even five or more pages of writing.

Looking Up the Size of a File

Whether you use the STAT utility of CP/M or the DIR utility of MS-DOS, each file name will be listed separately along with its size. Electronically stored data are measured in characters rather than by page length or inches. To make the numbers easier to read, the file sizes are sometimes shown in thousands of characters; for example, a 2,000-character file would be shown as 2K. But why would you want to know the size of a file anyway?

You'll certainly need to know a file's size if you're going to copy it to another disk. Copying just means getting one disk drive to read (play) and the other disk drive to write (record). This is very much like what you do when you "copy" a song from a phonograph record onto a cassette tape. Before beginning to copy a file from one disk to another, you first have to determine how large the file to be copied is and how much empty space there is on the disk to which you are copying it, so you'll be sure you have sufficient space to accomodate it. Of course, that's what we all do when we want to add something new to a limited space, but we do it intuitively.

Suppose you have bookshelf full of telephone books and you want to add a new one—the Chicago yellow pages. You would first look at how thick the Chicago telephone book is and how much empty space there is on the shelf. Those two pieces of information—the size of the file (phone book) and the amount of empty space on the shelf—are necessary before you can tell whether or not the new phone book will fit.

But you don't even think about that process—you just do it naturally. And that's one of the frustrating things about using a computer—you have to learn to think like a machine instead of a human being.

Erasing Files

If you want to discard a file, you simply ask the operating system to erase it by using the appropriate system utility. On CP/M computers the utility is called ERA (for erase); on MS-DOS computers the command is DEL (for delete); on Radio Shack computers, the command is KILL. (Not exactly friendly, is it?) Although the names for the function differ, all operating systems have an erase program.

In addition to typing in the name of the utility, you also have to type in the name of the file to be erased. Obviously, you wouldn't want the operating system to have to guess which file to kill. For example, when I want to erase the letter to my publisher that I had saved, I type the following: DEL PUBLISHR.LTR (if I used an MS-DOS machine). An exact match of the file name is required. If you misspell the name of the file you want erased, even by only one letter, there won't be an exact match and no erasing will take place. Instead, you'll see an error message displayed on the screen such as "NO SUCH FILE."

Renaming Files

Frequently, you will want to change the name you've given to a particular file. If you are dealing with manila files, you simply white-out the name and write in another. For example, suppose you have a file in which you keep monthly bills, called "Accounts Payable," but always think of it as "Bills" and keep looking in the wrong place in your file cabinet for it. You will probably put a new gummed label on the file folder and call it "Bills." All operating systems allow you to change the names on your computer files as well by providing you with a RENAME function as one of the utility programs.

Copying Files

Another utility function enables you to copy files. In CP/M systems the name of the function is PIP, in MS-DOS computers the function is called COPY. When using this function, you supply the name of the file you wish to copy (the original), the disk on which it is located, and instruct the operating system which disk to copy the file to (the **destination disk**). Each operating system has its own particular method for actually specifying the origin and destination disks, but let's not go into that here. That information is in the operating system manual that comes with every computer. Once you give the correct command, the copy function does the copying. That's all there is to it.

Safety in Backups

One of the first rules of using a computer is this: Make **backups**. A backup is a copy of your data that is kept on a separate disk. This copy should be stored in a safe place and never be used. It's sole purpose is to act as a safety net—to be used only if your original file is inadvertently lost or destroyed. If you make a mistake and accidently command the operating system to erase a particular file that you later realize is important or if your floppy disk gets ruined, you can always pull the backup copy out of its safekeeping spot and proceed.

Making a backup copy is quite easy. But many people overlook this obvious precaution. Still others make backup copies only occassionally. The rule to follow is make a backup every time any information in your file has changed. Some uses, such as mailing lists that are updated monthly, may require a backup once a month. Other uses, such as accounts payable, may dictate daily backups.

Many people ignore following these safety procedures until they have a catastrophe. It only takes one to make you become fanatical about making backups. After a few disasters of my own, I have become compulsive. My procedure is to make two backup copies. Since duplicating a disk takes only minutes, it's a small price to pay for peace of mind. I store one of the backups off site. That way even if there is a fire, my data will be unharmed. I know I can always buy a new computer, but there's no way to replace my precious data.

No matter how careful you are, it's inevitable that something inadvertent will happen to your file. When it does, you will be enormously thankful that you have a backup copy; if you didn't make a backup copy, you'll berate yourself for not having taken the few minutes necessary to ensure that your data were protected. Even if you think you are perfect and will never make such a mistake, disks do wear out. Remember, the disks are just mylar coated with iron oxide particles. As the read/write head rubs against the disk, some of the particles eventually start to fall off. When that happens, your data are worthless and you have to throw your disk away. If you have a copy of your information, you're safe.

It's also important to backup program files. In fact, that's the first thing you should do when you purchase a new software package—copy it. Then, if the disk gets ruined, you won't lose the money you spent to purchase the program. Although some software comes **copy protected**, which means you can't make a copy, most software is able to be copied. Often companies that supply copy-protected software will send you a backup copy if you mail in a special postcard that comes with the program when you buy it.

Although it's illegal to make copies of programs to give or sell to other people, the software companies expect you to make copies for yourself. They know that sooner or later something will happen to the original disk, especially if you use it every day, and they don't want to have to deal with it or you. Instead, they advise you to make a copy immediately. Put the original disk away in a safe spot and NEVER

USE IT! When your copy wears out in a year or so or you damage it by spilling coffee on it, merely pull out the original and make a new **working copy** to use.

Operating Systems' Command Structure

That's almost all there is to the operating system! It does those mundane jobs we take for granted when we use a regular filing cabinet, including showing the size of a particular file and how much empty space is left in the cabinet drawer, changing the name of a particular file folder, copying the contents of a file, and throwing out a file folder (erasing a file).

All the operating system programs that perform these functions are explained in the operating system manual. Regardless of which operating system you use, there is a common method that all the manuals have to denote how to activate the utility programs. This is called the **command structure**. Unfortunately, the manuals often don't bother to explain this structure. When you look up how to use a certain utility program, what you see is a bunch of confusing gobbledygook.

Personally, I found that one of the largest barriers to learning computers was trying to understand this standard command structure that was used to "explain" things in instruction manuals. Even though I knew what I wanted to do, the way that the instruction manual indicated to do it was indecipherable. Somehow they assume you're born knowing these things—just like they assume you innately know what boot means! To help you understand their assumptions, let's see how the functions we've described are set out in an instruction manual.

Square Brackets

If you look in a manual for a description of the DIR function, what you'll see is the rather cryptic explanation

DIR [d:filename]

The brackets mean that what's inside them is optional. If you just type DIR, leaving out the optional part, the utility assumes you want a listing of all the files on the disk.

Whenever you see brackets with a lowercase "filename," inside, you have the option of typing in the actual name of the file you're looking for. (Do not type in the brackets when you give the command.) That's another rule—lowercase letters get replaced with specific file names.

The advantage of typing in a specific file name is that it saves time if all you want is to know if a specific file is on that disk. This is particularly useful because the complete disk directory that's displayed on the screen may not be alphabetized, depending on the operating system.

Suppose you are using the utility to locate a specific file— say, CHAPTER 1. Rather than have to look through all the filenames that appear on the screen, you can type in DIR and the name of the file. You type

DIR CHAPTER 1

After you give the command, the operating system will acknowledge on the screen that CHAPTER 1 is on the disk or an error message will appear saying that CHAPTER 1 isn't on the disk. In essence you have a "smart" filing cabinet that you can ask this question: "Is the such-and-such file there?" And your "filing cabinet" will answer you.

Specifying the Disk Drive

Look back at the command structure for the directory utility. Inside the bracket you see a lowercase "d:" preceding the filename. A lowercase "d" followed by a colon (d:) always stands for disk drive. Each disk drive in your system has a name. Instead of referring to the disk drives as right-hand drive and left-hand drive or lower drive and upper drive, you use their one-character names. When I use my Kaypro, the first disk drive is called A and the second drive is called B. If I had a third drive it would be called C; a fourth would be called D; and so on.

All CP/M computers use this naming scheme. Computers that use the MS/DOS operating system, such as the IBM PC, also use this convention. Other operating systems have different schemes. For example, the Radio Shack computers call the first drive 0, the second 1, the third 2, and the fourth 3. Still another strategy is followed by Apple computers, which calls the first drive 1, the second 2, the third 3, and the fourth 4.

All operating systems have some assumptions built into them that are called **defaults**. For example, if you don't tell the operating system a disk on which to look for things, it defaults to (assumes) the first disk drive (whether it's A, or 0, or 1).

To check to see if the file called CHAPTER1 is on the disk in the first disk drive, you merely type

DIR CHAPTER1

To check to see if it's on a disk in any other disk drive, you need to designate which drive the operating system should look at. If you want to check the disk in the second disk drive, you type in

DIR B: CHAPTER1 or DIR 1:CHAPTER1 or
DIR 2:CHAPTER1

depending upon how your disk drives are identified. Notice that when you specify a particular disk drive, you have to follow it with a colon. The colon separates the drive you're specifying from the file name. If it weren't there, the operating system would think you were looking for the file BCHAPTER1.

In following command structure, you must do so *ex-*

actly as it's shown in the manual. Had you typed a space between the colon and the filename (DIR B: CHAPTER 1), the compare would not work because the CPU would look for a file with the first character being a *space*, the next being "C," and so on.

Another Example—Renaming Files

Remember, each operating system will have its own way of giving the operating system commands; however, the method of notation in all the manuals is the same. Now let's look at how the CP/M and MS-DOS instruction manuals tell you to rename a file.

Suppose you want to change a file named AP (accounts payable) to BILLS. In a CP/M operating system manual, the rename function is explained as follows:

REN d:newname = oldname

Now, following our rules, if your accounts payable disk were in drive B you would type

REN B:BILLS = AP

If you look at a directory after giving this rename command you'd see the file BILLS listed and the AP listing no longer there.

An MS-DOS operating system manual describes the rename function differently. It shows

REN d:oldname newname

Notice that the order of the file names is reversed and that instead of using an equal sign to separate the old and new file names, there is simply a space. To accomplish the same rename you would type

REN B:AP BILLS

Although the commands are cosmetically different in the two operating systems, you'll be able to use any computer if you are able to read, understand and correctly follow the command structure illustrated in the system's manual.

Formatting

There is one more utility program that is included as part of your operating system: It's the FORMAT utility. This program prepares new disks for use by your particular computer. Normally, when you buy blank disks they're purchased in a box of 10 disks. But these blank disks are not immediately useable because they don't have the sectors laid out on them. Remember our discussion about disks in Chapter 1? Disks have tracks and sectors that act as the parking spaces for your data. Until the parking spots are marked, data can't be stored on the disk.

Why don't the disks come with the sectors already marked on them? Imagine the inventory problem that a computer store or business supply center would encounter if it had to stock hundreds of different types of preformatted disks. There are countless disk formats, because every disk drive manufacturer has its own particular scheme for arranging the tracks and sectors on disks. One particular disk drive manufacturer may use 40 tracks on a disk; another may use 80. One manufacturer may use 16 sectors per disk; another may use 10.

To make matters easy for everyone, the disks are sold blank, without sector marks. You put the new disk into one of the disk drives and run the format program. This magnetically marks the sectors according to the specifications required for the disk drive your machine is using. Since the operating system comes tailored for your particular brand of equipment, the format program will only create formatted disks for use on your system.

Most format programs also test the disk for flaws, although some operating systems do flaw checking by using a separate program (called DDTEST). Whether these tests are included within the format utility or are provided separately, they verify that the disk is physically all right. They write a particular pattern out to the disk and then reread the disk to verify what the pattern is. If any particles of iron oxide have fallen off the disk, the pattern will not be recorded correctly, and an error message will be displayed notifying you that the disk is damaged.

Periodically, when I'm ready to erase and reuse an entire disk, I reformat it. By doing that and not simply erasing the files on the disk, I get a safety check performed, confirming that the disk hasn't worn out. Be careful about formatting disks that you've used. When you format a disk, you automatically erase all the information that was stored on it. Once a disk is erased, the data that were on it are gone forever.

CHAPTER 8

New Careers

When I first started teaching computer literacy, it made me somewhat nervous that students often had expectations for new careers they hoped would result from my course. I wanted to avoid raising false hopes about what career opportunities existed for newly computer-literate people because I certainly didn't want to accept tuition under false pretenses. I was particularly concerned about this for unemployed or underemployed students. When they asked about what they would get out of the class, I carefully refrained from the type of language you see on matchbook covers that promises complete life transformations for just a few dollars.

Actually, the benefits associated with computer literacy have amazed me. Students who took my two-day class that included hands-on training (they used computers for a total of 17 hours) and covered the material in this handbook, got incredible job offers. Students who were unemployed got jobs; others moved into new positions or opened their own businesses. At first I thought this was exceptional. But as more and more students wrote or called to tell me what had happened to them, I realized not only what a valuable tool computer literacy is but that employers appreciate it as a valuable job skill.

Some of you may already have ideas about how you can become more productive by putting the computer to work doing things that were both tiresome and time consuming when done manually. In addition to those benefits, there is a whole new range of jobs open to you, solely because you are now conversant in computerese. These new opportunities include jobs that specifically serve the computer industry it-self, as well as jobs that provide computer-assisted services. That's what the subject of this chapter is.

I am particularly enthusiastic about people combining their existing skills with computer technology. This means that rather than changing careers entirely, you can upgrade what you already know into an expanded career by figuring out the way for a computer to assist or enhance your work.

This chapter will look at some new careers that have emerged, many of which build upon nontechnical skills that many women already possess.

Technical Writing

Technical writing is a good example of how you can combine the nontechnical ability of writing clearly with basic computer literacy to create a well-paying job. The stereotype of the starving novelist is no accident—writing has always been an underpaid profession. But combined with computer literacy and labeled technical writing, it becomes a professional career.

Technical writing takes many forms. Writing documentation—the instruction manuals that come with the software and hardware—is one possibility. The skills associated with this field are clear, organized writing and a degree of comfort with computer jargon. There's a shortage of good documentation writers because these two traits don't normally go together. Most writers were English majors in college, carefully avoiding technical subjects. But technical writing doesn't require you to actually be a technician yourself. Generally, you only need to know enough to be able to interview the programmer who created the program and translate the information you receive into material that's comfortable for consumers to read.

The range of possibilities associated with technical writing is amazing. You might work for a software company that needs documentation materials designed for grade schoolers or middle-management executives. Or you might be part of a team composed of a project leader, a group of programmers, and yourself, the technical writer. This type of job is increasingly appearing at large corporations such as banks and insurance companies that have their own in-house data processing departments.

Of course, the opportunities for writing documentation for a computer company depend upon your location. Computer companies have traditionally been clustered around a few locations. The "silicon valleys" of northern California and the area around Boston, Massachussetts, have the highest concentrations of these companies. But an ever-increasing number of software companies is springing up all over the country—in Southern California; Seattle, Washington; Fort Worth, Texas; Atlanta, Georgia; Boca Raton, Florida; Minneapolis, Minnesota; the list goes on and on.

Getting your first job in this field is difficult. One way to break in is to create some samples of your work. It's easy enough to find programs to document. You can create a new 5- or 10-page description of some features of a well-known program. Of course, to do that you'll need access to a computer. If you don't have a computer or a friend who does, go down to the local computer store and ask to use theirs along with one of their software demonstration disks. Usually, it's a good idea to call the store first. My experience has been that as long as you don't want to come in during the lunch hour, the store personnel are quite accommodating. After all, it's to their advantage to have you become addicted to a machine that they're selling. Another way to get access is to volunteer at the local university or community college. Seek out the computer center manager and offer your services. You'll probably be viewed as a godsend and will likely get good feedback about your work.

Other types of computer-related writing include the new field of computer journalism and publicity-related jobs. To see the amazing variety of computer magazines and newspapers, go to a good newspaper/magazine store. Even going to a chain drugstore, such as Walgreens, is an eye-opener. Again, what you'll find is a wide range of target audiences— from magazines aimed at school children to highly technical and specialized publications for people in the computer industry. Since the field is booming, demand for articles is immense. If you have a good idea, write it up and send in a description of the concept to the magazine that seems most appropriate.

Public relations writing is also booming. Have you noticed the number of articles about computers appearing in your newspaper lately? Most likely they have appeared there because of the efforts of a publicist or communications specialist. This person writes press releases, new product announcements, profiles of key computer personnel, and so on. Once you've had something published, you increase your job possibilities.

Education

Are you a teacher looking for a change? Or were you trained in college to be a teacher, but because the days of baby boom ended, you never got to teach?

Teaching is directly transferrable to computer education, at a substantial increase in salary. As a trainer, you may be employed by a computer store to teach software packages to purchasers of new machines. Or you may be employed at large corporations to teach in-house classes to employees. Another job might be as a trainer for a software company that sells to particular industries. Often packages tailored to the needs of insurance agencies, real-estate agencies, insurance companies, hospitals, and banks cost $15,000 or more. When companies buy this type of software, they usually receive free on-site training.

The type of training required for these jobs depends on which type of job you want. Teaching in the school system usually means knowing programming—specifically BASIC.

Now that you've finished this handbook, you should be able to teach yourself programming quite easily. You will need a computer and a good book. A child's home computer is perfectly adequate for learning programming. You can get a computer that's suited to the single objective of learning BASIC for $200 or less. Look at the Commodore, Atari, and Radio Shack color computers. As long as the machine has BASIC in ROM, you don't even need a disk drive (unless you want to save the programs you've written). Of course, as I've pointed out in this handbook, don't expect to be able to do much else with a home computer other than learning BASIC programming.

Other training jobs don't require any programming ability at all. What they require is an in-depth knowledge of one or two popular software packages and a willingness on your part to learn more. To learn a package, sit down with a computer, a readable instruction book, and the software. As described above, most computer stores are gracious about letting you sit and work at one of their machines. But try not to let them help you too much. You need to do the work yourself. Only ask for help when you're really stuck. As a woman, you will most likely be offered more help than is good for you. Although I appreciate such chivalry at gas stations, in computer stores I cheerfully decline. It's only by working out the details of a software package on your own that you have the confidence to tackle another package, and another, and yet another. When a computer store hires a trainer, it needs someone who will keep up with the new software as it is developed.

Some training jobs require an in-depth knowledge of a particular industry and may require absolutely no knowledge of computers. Software companies have realized that it's relatively easy to teach a new trainer about computers, but it's almost impossible to teach the ins and outs of a particular industry. If you've worked in one of the paper-based industries such as insurance or banking, particularly in a narrow specialty such as trust accounting or medical bookkeeping, you can bring that experience to your new computer training job.

User-Interface Specialist

Doesn't **user interface** sound impressive? A user interface specialist is a person who feels comfortable as the link between the "techies" (the computer programmers) and the people who will ultimately use their new program (the **end users**). Classically, neither of these groups has been able to talk to each other. This new job was developed to facilitate communication.

This job is needed because traditionally programmers are immersed in talking to machines, not people. After a while, whatever interpersonal skills they had decline. They also forget how to express ideas in nontechnical language. At the other end of the conversation is the end user. Typically, this person gets anxious whenever computer jargon starts creeping into the conversation. The end user may know what she wants the computer to do, but she may have a hard time translating her requirements into language that the computer specialist understands. That's where the user-interface specialist comes in.

In addition to liking people and being computer literate, any skills you have in a particular industry will help you get this type of job. In many situations, specific experience in the particular industry is a necessity, since you have to be literate in the end user's field as well. That means if you are a user-interface specialist in a large bank, you also need to know banking terminology; in the insurance industry you need to know insurance vocabulary; and so on.

Quality Control

Quality control means making sure that no one goofed. This job is often a subspecialty associated with documentation writing. Your job is to follow the directions in the documentation that someone else wrote and see if everything works. If you get stuck and can't understand them, the end user probably won't either. The only job-related skills you need for this job are to be computer literate and to be reasonably comfortable with computer jargon.

Start a Business

Starting a business has traditionally meant that you needed a large amount of capital behind you. But, with the cost of computers as low as it is now, a computer-based business actually requires very little capital. A $2,000 computer purchased by credit card will cost you about $75 a month.

We are moving into an information-based economy and information is now a commodity, something to be bought and sold. Many new businesses have been started by individuals who identified a particular type of information that people wanted and created a database to provide it. Often these entrepreneurs didn't do anything more than take public information, such as that scattered in many different reference books, and make it more accessible.

One way to structure a business is to build a database and charge for searches. For example, one enterprising college student researched college scholarship programs. He created a database with one record for every scholarship. Associated fields were the criteria necessary for awarding the grant, such as major, income, applicant's age, religion, and so on. He built the database by doing library research. He then opened his electronic file to others for a fee. To use the database, an applicant fills out a form listing relevant data such as field of study and parents' income. Then a search is run, and entries from the database that match the applicant's characteristics are printed. The cost for such a search is about $35.

Depending upon the nature of the database, searches may be much more expensive. Usually the fees for a search are directly related to how arcane the information to be searched is. For example, at UCLA the Brain Information Service has a team of medical librarians cataloging articles that relate to medical research on brains. The librarians create an abstract, consisting of a short description of the article and a list of **keywords** that can be used to trigger a match when a search is done. Medical researchers from all over the world send in their search requests, listing the type of articles they want to know about. They can link the criteria using connectors such as AND and OR. For example, a researcher might want to know only about material pertaining to the hypothalamus AND vitamin B. Annual subscriptions to this database are very expensive.

Obviously, the amount of capital to start such a highly specialized search service is substantial. The medical researchers in the example above needed a salary, and no searches could be sold for quite a few years until the data were cataloged. At the other extreme, the scholarship service only required an investment in time and the purchase of an Apple II with a hard disk and a database program.

Another type of database service requires absolutely no research, only typing. Computerizing standard reference books is becoming increasingly popular. If the reference work is copyrighted, license fees will probably be required. But a huge amount of reference material is public information provided by the U.S. government. For example, one company has computerized the U.S. Congress Handbook. This directory includes bibliographic information on members of Congress, committees and subcommittees, legislative aides, press secretaries, and more. Instead of selling searches, the company provides the information in database form on disks. Then purchasers can produce their own mailing lists, selecting names according to various criteria. The cost to buy this database is $295.

Still another database service creates information where no source exists. For example, many people have begun to swap houses for vacations. A database such as this may charge for listings as well as searches. Other creative database services include carpooling services, baby sitting exchanges, and tennis partner matching services.

Connecting such databases to modems provides still another extension to these services. For example, in California, high-tech jobs in the computer industry can be searched by job seekers from their homes using their own terminals. Searches are free, but prospective employers are charged a fee to list the job opening.

Still another application is simply maintaining someone else's database. For example, small associations and clubs may not find it cost effective to maintain their mailing lists. Instead, they contract with a private company to do it for them. This type of service usually charges for the initial data entry and then for each correction, insertion, or deletion from the list. Mailing lists in label form are provided and charged for at a per-name basis.

Word Processing

Other businesses offer word processing services. The only capital required for such a business is money to buy a computer, a word processing program, and a letter-quality printer. Purchasing a Kaypro, which includes word processing software (WordStar) in its purchase price, means the total cost is under $2,000.

One commercial application of word processing is a résumé service. Clients have their résumés prepared for them and permanently stored on disk, ready to be revised or updated (for a fee) each time the client applies for a different type of job or needs an update.

Another business that's popular around university towns

is dissertation and thesis preparation services. Since most graduate students have to do a series of revisions, this service types the first draft of the paper, stores it, and then does revisions.

Your own creativity is your limiting factor in trying to put a computer to use to create a new business. For example, one enterprising person has created a package of 100 business letters for sale on a floppy disk. These are special-purpose letters such as collection letters that dun for past-due payments and rejection letters to jobseekers. The letters were word processed using WordStar, and that was all the investment required.

Spreadsheet Services

For the bookkeeper who wants to leave a salaried job and begin freelancing, a spreadsheet-based business may be perfect. It is much more cost effective to enter expenditures into a spreadsheet than to do the work using a calculator. The spreadsheet will be able to generate monthly statements, quarterly and annual reports, cash flow projections, and other financial management reports for which businesses pay good money.

Increasingly, the analysis necessary to transform the spreadsheet into these forms has been done by experts—CPAs and financial analysts. The appropriate formulas have been sold in the form of templates that allow the information on the monthly expense ledger to get transformed into specific financial reports. Of course, if you are a financial planner with some specialized approach, you can sell your own methodology in the form of these templates.

Whether or not you use your newfound computer literacy to start a new business, get a better job, or have a useful new skill for managing information in the years to come, it's important to remember that whatever you want to do with this new technology *you can do!* Although the overwhelming cultural message is that computers are inaccessible (particularly to women), I trust you will now have the courage and self-confidence to take what you've learned in this handbook and go out and teach yourself whatever else you want to know, with the certainty that your goals are attainable.

Technical Notes

These technical notes contain advanced material that is unnecessary to attain computer literacy. They are included for those who already know something about computers and desire a more technical level of detail. As such they assume an advanced level of understanding—the computer terms used here are not always defined. Beginners should not even attempt to read these footnotes as they read the book. You won't ever need to know any of this material unless you plan on a career in either hardware or software design.

1. The word *byte* is technically defined as the *representation* of a character. This definition may have been created in order to include the representation of a command as well. For example, in an 8-bit byte (used by personal computers as well as IBM mainframes) a command such as "add" is also represented using 8 bits.

2. Actually, the CPU has a few other capabilities, such as the ability to move information from memory (RAM) into the CPU or move it back to RAM. The CPU also has the ability to shift the position of a byte within its registers. This permits the machine language program to multiply or divide by factors of 2 in the same way that moving a decimal point to the right or left allows people to multiply or divide by powers of 10. This capability also enables programmers to ascertain the setting of a particular bit.

3. The time it takes to execute an instruction depends not only on the actual time required to perform the task but also on the time it takes to move information in and out of the CPU. For this

reason, computers with the same cycle time but different size buses will perform an instruction in a different time interval.

In addition, a task is composed of two types of functions—computation and communication with the peripherals. Having a faster CPU will only speed completion of a task relative to the percentage of computation performed.

In more sophisticated computers, controllers free up the CPU so that once a command is passed to a peripheral's controller, the CPU can continue processing. In less sophisticated computers (such as the Sinclair/Timex), peripherals are controlled by cycle-stealing. Similarly, the type of interrupt structure, queueing structure, and so on, ultimately determines throughput.

4. Actually, K really means 1,024. However, there's really no reason to remember this strict technical distinction since all references to file size, program length, and memory are made in terms of K, thus canceling out the factor of 1024. For example, if a program requires 64K RAM and you have a 64K computer, you know you have sufficient memory without having to know that you actually have 64 times 1024 RAM.

5. Some disk drives achieve higher density by recording more data in a sector. Sometimes this is achieved by the actual bit structure. When stop/start bits are recorded, this lowers densities (but ensures compatibility across different manufacturers' drives). Sometimes the higher density is achieved by recording more densely (more bits per inch).

6. Strictly speaking, if you have a lot of RAM, more of your data can be transferred to memory in one "physical" read. This means fewer reads are required. Since transferring data from the disk to memory is a slow procedure, processing is speeded up when fewer disk accesses are required.

7. Memory must hold the operating system as well, but we'll get to that in Chapter 3.

8. There are various ways to represent the letter A. 01000001 is the ASCII representation. ASCII stands for American Standard Communication and Information Interchange.

9. Usually chips are rated in terms of kilobits. To simplify things, I have been referring to them in their equivalent kilobyte size. The 8K RAM chips are strictly speaking 64 kilobit chips (ab-

breviated 64Kb), and IBM's experimental 128K chip is more accurately called a 1024Kb (1024 kilobit) or IM chip.

Figure 1-16 has been simplified as well. Since each memory mailbox has 8 bits, it requires 8 wires as a bus. Depending upon the system architecture, there may be eight wires going to each RAM chip, or a memory mailbox may be composed of 1 bit in each of eight RAM chips. The latter approach is more common. This level of detail is only necessary for hardware designers. A more realistic explanation of parallel communications is provided in Chapter 3.

10. When sending ASCII characters, an approximate value of characters per second can be obtained by dividing baud by 10. For example, 300 baud sends approximately 30 characters per second. Sending a page of information (3,000 characters) would take 100 seconds or almost 2 minutes. Sending the same page at 1200 baud would only take 25 seconds.

The reason that a rating of characters per second isn't used is that there are various communications protocols. Bits are inserted at the beginning and end of each character (stop and start bits). Depending upon the convention used, different numbers of bits must be sent in order to send one character.

11. However, by adding many digits at once instead of one column at a time (as you do manually), processing time can be significantly speeded up. For this reason, math coprocessors are available as an optional feature for many brands of microcomputers. The registers of these math coprocessors are 80 bits long.

12. "Line up the decimal points" refers to the entire FD section of the COBOL program. This ability to define entire page formats is one of the features of the COBOL language that clearly distinguishes it from FORTRAN.

13. Strictly speaking, "translated" means that the entire program isn't converted to machine language before execution. Instead, each line is converted at run time. This is a much slower approach than compiling. In compilation the entire program is converted and stored in machine language. Thus, the overhead of conversion is incurred only once, although the program may be used many times.

14. Actually there are some reversible methods for "freezing"

molecules. For example, EPROM can be erased through the use of ultraviolet lights.

15. Since these system programs drive the peripherals, they are known as **drivers**. There are other routines included in the operating system; these are described in Chapter 7.

16. Some computers (usually home computers) have the operating system in ROM so no disk drives are required. The IBM PC is one of the few business computers that has a ROM chip with the operating system. It also has BASIC in ROM. When you turn on an IBM PC (without a boot disk in the disk drive) it will automatically put you into BASIC. IBM probably designed its computer with this capability so its PC could be sold to schools. If schools bought IBM PC computers without disk drives, they could still teach BASIC programming.

17. The disk must be placed in the primary drive: drive A in CP/M and MS-DOS, drive 1 in Apple DOS, or drive 0 in TRS-DOS computers.

18. Actually the on button isn't wired to the disk drive. The instructions that initiate loading of the operating system are stored in a boot ROM. This ROM is activated by the on switch.

19. The name of the section of the operating system that gets tailored for particular peripherals is called the **BIOS** for **Basic Input Output System.**

20. Actually, there is a version of CP/M that works on 16-bit computers. This operating system, CP/M-86, never caught on with software developers and there are only a few programs that require it. Clearly, MS-DOS is the standard for 16-bit computers.

21. Actually, due to memory requirements, software developers write Macintosh programs on the Apple Lisa. Still, no software development could occur until both the Lisa (machine) and Lisa/Macintosh language compilers were available.

22. Lotus Development Company supplies the Lotus program with drivers for each of the major brands of IBM-compatible computers. The user then specifies which type of computer she is using and thus installs Lotus for her particular brand of computer.

Index

This index was prepared with the help of Starindex, an indexing program from MicroPro Corporation that is a companion to their Wordstar wordprocessing program.

PLUME

EXECUTIVE'S CHOICE

CREATIVE COMPUTING

☐ **BASIC PRIMER for the IBM® PC and XT by Bernd Enders and Bob Petersen.** An exceptionally easy-to-follow entry into BASIC Programming that also serves as a comprehensive reference guide for the advanced user. Includes thorough coverage of all IBM BASIC features; color graphics, sound, disk access, and floating point. (261961—$24.95)

☐ **DOS PRIMER FOR THE IBM® PC and XT by Mitchell Waite, John Angermeyer, and Mark Noble.** An easy-to-understand guide to IBM's disk operating system, versions 1.1 and 2.0, which explains—from the ground up—what a DOS does and how to use it. Also covered are advanced topics such as the fixed disk, tree-structured directories, and redirection. (260450—$16.95)

☐ **BLUEBOOK OF ASSEMBLY ROUTINES for the IBM® PC and XT by Christopher L. Morgan.** A collection of expertly written "cookbook" routines that can be plugged in and used in any BASIC, Pascal, or assembly language program. Included are graphics, sound, and arithmetic conversions. Get the speed and power of assembly language in your program, even if you don't know the language! (254981—$19.95)

☐ **THE COMPUTER LOG: The Best Thing Next to Your Computer by Howard Hillman.** If you use a computer, you need *The Computer Log*. The carefully planned logs and directories in this oversized organizer will help you find those crucial numbers, codes and lists you often need at a moment's notice as you work at your computer. Some features: task and activity log, on-line database directory, electronic mail service directory, repair service directory, software inventory, vendor directory, and much, much more. Finally, one place for all your information! (256488—$12.95)

☐ **MASTERING SIGHT AND SOUND ON THE IBM PCjr. Kent Porter.** A gentle but thorough guide to BASIC programming for music and graphics, it covers BASIC commands, jr's operation system, and the new super-powered graphics software that takes jr beyond Apple's Macintosh. (254906—$9.95)

Prices slightly higher in Canada.

Buy them at your local bookstore or use this convenient coupon for ordering.

NEW AMERICAN LIBRARY
P.O. Box 999, Bergenfield, New Jersey 07621

Please send me the PLUME BOOKS I have checked above. I am enclosing $_____ (please add $1.50 to this order to cover postage and handling). Send check or money order—no cash or C.O.D.'s. Price and numbers are subject to change without notice.

Name_____

Address_____

City_____ State _____ Zip Code _____

Allow 4-6 weeks for delivery. This offer is subject to withdrawal without notice.